GOATKEEPING
A Guide to Getting Started

by
Iris Staniland

THORSONS PUBLISHERS LIMITED
Wellingborough, Northamptonshire

First published 1979
Second Impression 1980
Third Impression 1982
Fourth Impression 1984

ISBN 0 7225 0504 3 (hardback)
ISBN 0 7225 0503 5 (paperback)

Printed and bound in Great Britain.

CONTENTS

I should like to express thanks to my family and friends who have made the writing of this book possible by their many hours of help around the farm and the provision of additional literature and photographs.

1

HOW TO START

In the play *A Ring of Roses* we are told the story of the village of Eyam, when the arrival of a basket of second-hand clothes from London results in the majority of the people dying of the black death. In an effort to escape, one old man suggests he takes to the hills to live. 'If I take a few hens and a goat, I'll be all right,' he says. Today, goatkeepers, and potential goatkeepers, are endeavouring to escape, too. They are escaping from the very high cost of living by producing as much of their own food as possible. Goat products will help a great deal with this, but goatkeeping must be done efficiently, and the goatkeeper must have done some homework beforehand.

How and Where to Start

The British Goat Society is probably the first source of information that springs to mind. If you contact them at Bury St Edmunds they will send you a list of their publications, the most up-to-date list of addresses of all affiliated clubs, details of their work, ideals, and subscription rates.

However, to start with, your local club will probably be of even greater value to you, simply because it is local. Join your local club even before you have your own goats. Get to know other goatkeepers in your area; you may even find there is one on your doorstep. The secretaries of these clubs are invariably busy people but they always seem to find the time to help beginner goatkeepers, or are able to put beginners in touch with someone who can help.

Many clubs run Buying Groups and these are very useful, not just from the financial point of view, but because many of the items needed daily can only be purchased in large quantities. If your pocket is deep enough and your house big enough to store goods, this could be no problem, but many things have a comparatively short shelf life which could result in the bulk of an item being thrown away after, say, three months. Milk substitute and some sterilizing agents are prime examples.

What are you going to do with your goats when you go on holiday? Can a neighbour spare the time to milk them twice each day as well as coping with the feeding? Will the beginnings of illness be recognized? Most clubs publish a magazine three or four times a year, and members often advertise their willingness to board goats. It is so much better to be able to relax and enjoy your holiday knowing your goats are in capable hands than to keep wondering if all is well at home. Friendship, information and help are just some of the many advantages of joining a club. The whole family is welcome at the get-togethers which are usually held at different places on each occasion so that it is possible to attend at least once each year. At a Spring meeting it is not unusual to see four-legged and two-legged kids being given their respective bottles.

Books are, of course, valuable sources of information, and many magazines are running series of articles on the subject too. Never be tempted to think you know all there is to know about goats. Keep your eyes, ears and mind open to learn at all times, and you'll find a whole new world opening before you.

Buying Your Goats
It is very seldom a good idea to buy in answer to an

advertisement in a newspaper. So often you see something like: 'Pedigree, white nanny goat for sale'. On inspecting the goat and deciding to buy, you ask for her registration card and are then told, 'I'm afraid I've mislaid it, but she is a pedigree'. Until you have proof of this and the card in your hands, that animal is no more a pedigree than last Sunday's joint. In fact, you probably never will receive the card, so be warned. Buy from a reliable source. Most fellow club members are caring, reliable people so buy from one of them if you can. Your club secretary should have a list of goats for sale. Before you finally decide to buy a specific goat, do go and see it, making an appointment beforehand; try to see two or three others of the same family and so build up a picture in your mind of what your future herd could look like. Never be afraid to ask questions about her or her relations. A goatkeeper is more likely to be willing to sell you a goat if it is obvious it is going to a caring, if inexperienced, home.

Look to see whether 'your' goat has a ladylike appearance, with a long, thin, milky neck. Are her eyes bright and does she look in good health generally? Is her coat shiny? Is she interested in everything that is going on around her? Has she a good deep body, indicating she is able to eat a large amount of roughage? Has she a gently sloping rump or a steep one? She could have difficult kiddings if she's too steep there. Also, her butterfats tend to be low if she has a steep rump. Is her udder well attached forward and high at the back, or is it just like a long sausage? If it is long and pendulous it will certainly get very scratched and easily knocked, so you will be needing the vet to sew her up from time to time. Look at her in profile. Does she have a leg at each corner or are all her feet centred under her tummy? If they are, she is probably minerally deficient. When you are an experienced goatkeeper you can deal with this, but don't touch her with a barge pole now. Get someone to walk her

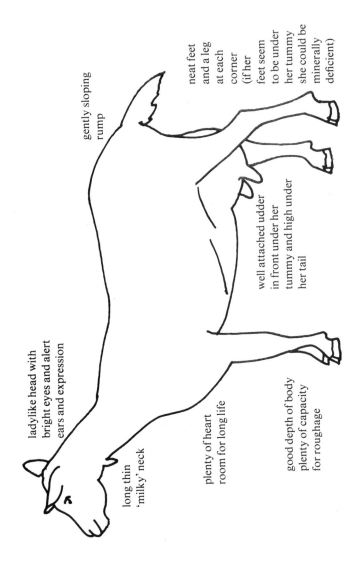

gently sloping rump

neat feet and a leg at each corner (if her feet seem to be under her tummy she could be minerally deficient)

well attached udder in front under her tummy and high under her tail

ladylike head with bright eyes and alert ears and expression

long thin 'milky' neck

plenty of heart room for long life

good depth of body plenty of capacity for roughage

around so that you can check whether or not she's cow-hocked. If her hocks turn inwards every time she takes a step forward, she will knock her udder.

Once you have decided to buy a goat, try to arrange a couple more visits before you actually bring her home so that you can milk her under the eye of the previous owner. She will settle in her new home much better if you do this. It is also a good idea to buy from her previous owner sufficient concentrates for, say, two or three feeds. You can then gradually change over her food to whatever you decide to use without upsetting her tummy.

A Companion

Goats are gregarious animals and are always much happier if they have a companion, preferably of the same species. If this does not happen, you will find that she will try to put *you* in this position. This is no doubt good for one's ego but not always very convenient! If she is on her own, she will call more frequently and keep trying to follow you wherever you go, like Mary and her little lamb. So for your own sake, as well as for your goat, start goatkeeping with two goats, never one.

Choosing a Male

Once you have six milkers you should buy your own stud males but until then let someone else do the work. Whilst most adult males are gentle, loveable animals, they are also very smelly during the rut, or mating season, and enjoy decorating their front legs and beards with urine as well. They need careful, patient training when young to avoid them being labelled vicious. A male takes five years to reach maturity, as against two years for a female, and weighs up to 5 cwt (254kg) at this age. They need separate housing, well away from the general public, and strong, high fencing.

Remember, the milk comes from the male line so do investigate each male's pedigree carefully. Always use a pedigree male so that you can do this. Try to visit his sisters, his dam (mother) and his granddam. Check milk yields. Ask questions. The owner of a good stud goat will be happy to answer your queries. *Never* use the 'billy down the road' just because he is near. Only use him if he meets all your requirements, and is a pedigree animal.

Toggenburg male (Tanatside Carlo)

Your club secretary will have a list of stud goats in your area and the British Goat Society also publish a list.

If you choose a young male, book a service in advance to avoid being disappointed. Wise stud-goat owners limit the amount of work he does in his first, and often second, working year. Always check with the stud-goat owners that it is

Saanen buckling (Wraxall Jasper)

convenient. Never just arrive on the off-chance. If the male has already been booked to work twice in one day you may be asked to wait until the following day or even until the next heat period. To book ahead avoids wasted journeys.

'In Season' Signs

The first indication you will have that your goat is in season will probably be when she starts to call a great deal. You will then notice her tail is wagging in little short bursts of energy. If you lift her tail you will see that her vulva, that tiny 'V' shaped piece of skin, is very pink and swollen and there is a white or clear mucus dripping from it. A milker's yield sometimes drops during this time too and, she is inclined to be fidgety to milk. Heat periods can last from twelve hours only up to three days so *know your goat*. You may have to act quickly. She comes into season every three weeks from approximately September to April, depending on the weather.

So long as you see three of these signs you can feel pretty certain she is in season. However, just to be quite sure, just before you leave to take her to the male put your hand on the goat's rump and gently bear down. If she stiffens her back legs she's ready for service. If she sags and walks away, you are either too late or need to wait a few hours, possibly until the next day, depending on her usual length of heat.

Mating

When visiting the male during the rut wear old clothes that can easily be washed or wiped down, and strong boots or shoes, never sandals. Some males have musk glands on their sides as well as near their horn buds. Leave your goat in the car or trailer until you have checked the stud-goat owner's procedure. (It is illegal to put your goat in the boot of the car.) Some stud-goat owners bring the male to you, others put your goat in his pen (not the best method as she is trespassing on his territory), and others let them meet outside his yard. Some like you to hold her, others will hold her themselves and let him run free. If you are asked to hold her be prepared for her to do the highland fling if she's a young goat, but always try to keep one of your knees in front of her chest when he lunges. A male goat does not waste time. If your goat is ready he works straight away, so be alert and stay calm. It often takes longer for you to ring the door bell than for him to cover her!

If your goat is registered, remember to take the card with you and remember to collect the service certificate and the card before you leave. If you have travelled far, get her served a second time before you leave. Although your goat has been mated she will continue this heat until its normal end, but in three weeks' time, check she is not in season again, and check again six weeks after mating. The gestation period is 150 days or five months.

2

GENERAL MANAGEMENT

Having decided to take the big step of keeping goats, you will
have to think very carefully about keeping them within your
own boundaries. Keeping fences and hedges in good condition is
therefore a priority. They should be inspected regularly, so that
you find the gaps before the goats do. A hole or gap in a fence is
a two-way thing; not only can your own stock stray, but other
animals can also find their way in and perhaps devastate a crop
of young cabbage plants or a future crop of hay. A great deal of
ill will is caused if this happens because it could have been
avoided with forethought.

It is therefore a good idea to have a safe area so that when
you do have to leave the animals for two or three hours, or
longer, you can relax without constantly worrying that they are
all right. This can be expensive but the peace of mind it brings is
well worth it.

Types of Fencing

Chainlink fencing is the only real answer but a 3-inch (7.5cm)
mesh is quite adequate. It is false economy to get anything less
than $\frac{1}{8}$-inch (3mm) thick because the animals rub themselves
against it, particularly in the Spring. Aim for a 4 foot 6 inch
(1.5m) height over all. Should you decide to put the fence on top
of a low wall, remember to leave a few small drainage holes at
the bottom.

If the goats are lucky enough to have a paddock as well as a safe area, the fencing must be strong enough here too, but because it is a larger area the fencing need not be quite so high say, 3 foot 6 inches (105cm).

Sheep or pig fencing can be used to fill gaps in hedges, and as a fence, but this can only be purchased up to 32 inches (80cm) high so two, or even three, straining wires at 6-inch (15cm) intervals above the sheep fencing are essential. Droppers, which link the straining wires vertically, are needed to keep the wires taut.

Wire netting, as used for chicken runs, is just not worth considering for confining the goats, and never use *barbed wire*. A ripped udder is not a pleasant sight and can be difficult to deal with.

Concrete posts are sometimes difficult to obtain but many large agricultural estates and some Forestry Commission depots sell the tree thinnings which make excellent fencing posts at a reasonable price. Do try to soak all posts in preservative before erecting them. Alternatively, buy posts already treated. They usually have a guaranteed life of many years.

Electric fencing units are useful for awkward shaped gardens or corners of fields, which can then be cleared by the goats. Three or four wires are needed to contain the goats adequately and it is a good idea to make the goats actually touch the wire just once when the unit has been switched on. When it is dry underfoot a hoof or horn may just be brushed against the wire. The current can then be switched off for quite long periods, because the goats know the wires may 'bite'. A steep hillside can also be fenced with this unit.

Tethering

Tethering is hard work for the goatkeeper if it is done properly, but it is not the recommended method of controlling the goats. A mobile shelter helps a great deal but very few tethered goats enjoy such luxury. Where no shelter is available the goatkeeper has to be alert for changes in the weather, and possibly has to put the goat in and then out three or four times in as many hours.

There are times when it is necessary to tether a goat for a specific purpose such as at a show where space could be limited, so do get them used to being on a tether for short periods. A wide collar is better than a narrow one as some goats have a blood vessel or nerve very near the surface on the right side of their neck and when pressure is put on this spot the goat collapses. Her legs seem to go in all directions at once. As soon as the pressure is released she will get to her feet and appear to be none the worse for the experience but it can be frightening to her owner the first time it is seen.

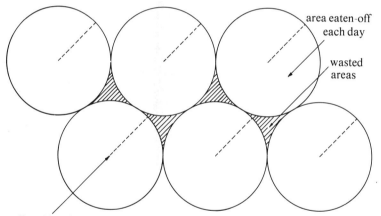

area eaten-off each day

wasted areas

spike with swivel

Tethering with single stake

Have three swivels in her chain and there is then less likelihood of her getting tied up. She can so easily hang herself if she is only on a rope. One of the dangers of tethering is that should the tether suddenly be pulled taut you, or a child, could be tripped up or even flung backwards and be concussed. It *has* happened.

A running tether is better than a chain on a stake. This need only be 3 feet (1m) long over a 15-foot (5m) run and is much more economical on the land as none need be wasted.

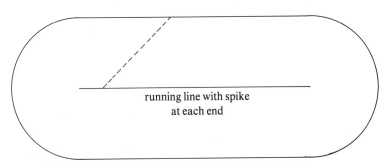

running line with spike
at each end

Tethering with running lead

Gates

Gates must be as high as the fences but this often means an extra bar or two, or strand of wire, must be added above the existing gate. If wire is used, support it with droppers at approximately 3 foot (1m) intervals, or it will sag.

Gate catches and door latches are no good unless they are strong and reliable. Aim for ones that need two movements to get them undone. Goats are renowned for their ability to undo bolts without a safety catch, or chew through string or rope, so do not put temptation in their way. A spring-loaded bolt is satisfactory while a lorry tail-board 'antiluce' catch is ideal for internal doors, especially if children are helping to shut them because there is a ratchet on the catch. This is an advantage if the youngster is unable to close the door fully.

Garden gate catches are often brittle and padlocks are a mixed blessing. Leave them both alone.

Housing

Your goats have few requirements, and once these have been met it is time to think of your own comfort. You will spend many hours each year amongst your goats, of necessity as well as of choice. They only need to have dry backs and be away from draughts, although their housing should be well ventilated. You need to be able to stand upright and she needs to be able to turn round. As far as she is concerned, a chicken house, coal shed or outside loo will be comfortable, but her attendant may not be too happy working in any of them.

Whether you decide to convert an existing building or build a new goat husbandry unit, a lot of thoughtful planning must go into it. A complete unit will consist of a pen, or pens, milking parlour and dairy, as well as a food-store and yard, but before making any major alterations to your premises, check with the local authority as to whether planning permission is needed. Agricultural buildings of this nature are usually exempt but sometimes one has to comply with building regulations. A friendly chat with the local planning office could be of value.

Fitting in this type of work with an already busy schedule can be difficult. Spread the programme over five years and you will then be sure you know which type of unit will work for you. Many people aim for just the shell of a building and yard for the first year. It will not be perfect but it could be adequate, and the cost is being spread over a period.

When you reach the stage of installing pens, these should be approximately 5 foot (1.6m) by 6 foot (2m) in size to enable the goat to move around freely, especially during the winter months when she cannot go out so much. Pen floors should have a gentle slope on them and be insulated. A cheap way to insulate the floor is to put down a thin skim of concrete to give a flat surface, up-end jam jars, beer cans, egg boxes, old double-roman type tiles, and so on – anything, in fact, that will give you a pocket of air – then cover the whole with another layer of concrete and you will have a floor that is never dangerously cold.

Alternatively, encourage the goats to lie well above the floor by providing a deep layer of straw, or a wooden bench. Benches, about 1 foot (30cm) high, need to be no more than 3 feet (1m) long or they will be difficult to move and clean. Wooden slats are another way of keeping them off a cold floor, but the slats must be quite close together or the goats could trap their hooves or damage their udders when getting up.

Strips of insulating material sandwiched between the roof and ceiling are one way of avoiding condensation dripping from the roof on frosty mornings. Another is to use a bitumen-covered fibre roofing sheet. These sheets can be bought in colours to blend with an existing dwelling or barn roof and are particularly easy for the ladies of the family to handle as they only weigh 15 lb (7kg) each. The only tool you will need is an ordinary wood saw for cutting around awkward shapes or corners.

Goats will spend many contented hours cudding by a door or window if they have a block to stand their front feet on. This also discourages them from using a bracing strip on a door as a perch!

Feeding Racks

Hay racks should be sited so that they can be filled without the attendant having to open pen doors each time. The front, or the whole of the rack, can be made from mesh, $2\frac{1}{4}$-inch (56mm) or the back and sides can be made from any scrap of wood that is strong enough. If the mesh is larger, a great deal of the hay will be shaken on to the floor and wasted. A trough placed under the rack will catch some, and this can be returned to the rack later. All hay racks should have a lid that can be clipped down or the goats will find it quicker to help themselves from the top, resulting in an even greater loss of hay.

Another time-saving idea is to fix bucket rings on the outside of the pens at fractionally above tail height. The goats then put their heads through the appropriate aperture to reach their food or water, and the buckets can be replenished quickly and easily.

Mineral lick holders should be included in the furnishings, but these can be placed one between each pair of pens and shared.

Dividing the Pens

All pen doors should be 3 feet (1m) wide, so that a wheelbarrow, or even two pregnant dams-to-be can pass through with ease and without causing any damage.

Metal bars are one way of dividing the pens but do limit the spacing of the bars to 4-inch (10cm) centres for adults and $2\frac{1}{2}$-inch (6cm) centres for kids. Be accurate with the measurements. It can be a time consuming business getting a head back through the bars if the goat prefers the view from the wrong side of the partition.

If space permits make any corridors or passages 4 feet (1.3m) wide.

Many people make a partition at the end of the passage and use this area as a milking parlour. Provided it is easy to keep clean this is fine. A milking bench takes the backache out of milking but it must be strong enough to carry a 150-lb (68kg) goat plus the weight of the heaviest member of the family who will be doing the job. The height of the bench will vary from 10 inches (25cm) to as much as 14 inches (35cm) depending on the most comfortable height for the person doing the milking. It will have to be scrubbed regularly but it must not be slippery. Once a goat has slipped it could take a while to restore her confidence.

You will find your goats will enjoy being milked, especially if they have a raised block to put their front feet on. Fix a ring, through which you can clip a chain and clip, to the wall at the side of the bench. When you first start milking your own goats, you may be rather slow and they may get bored and try to wander back to their pen. A tie acts as a deterrent. Goatlings are particularly prone to this form of activity but it is a help if the goats have always associated the bench with pleasant things. Encourage them to get up there for hoof-trimming and grooming and give them a small titbit afterwards.

A bucket ring is useful on the front of the bench for occasional or regular feeding but make all bucket rings as strong as possible. When they are not being used to hold buckets they occasionally get used for scratching between the horns or horn buds and this does the rings no good at all.

The Milking Parlour and Dairy

A sloping, concrete floor is an advantage in the milking parlour, as it is in the dairy. A north facing dairy is best, but the main thing is to get your dairy work away from the kitchen as soon as possible. A storage cupboard for milk filters, cartons, and so on, that can be used as a working surface is ideal. A large plastic bowl will complete the initial equipment needed in the dairy. In time, a water supply and an electric point for a refrigerator can be added. If you are doing the installations yourself, alkathene tubing is easy to use but check the grade needed before buying it and remember to position all taps and electric switches well out of the reach of the goats. They can stretch their necks a very long way!

The Food Store

Try to position your food store so that there are at least two doors or gates between it and the goats. A goat that has been able to raid a bin could be very ill indeed. Put a decent catch or lock on the door. Dustbins with well-fitting lids make good storage bins. Any bags of food not in bins should be kept off the floor. A wet Wellington boot quickly damages a bag and wastes the contents. It is better to stand a bag on two bricks than on the floor.

Make your food store big enough to keep buckets, broom, medicine chest, and current bales of hay and straw in. Fit a row of hooks for collars and leads. Have a torch that works there summer and winter in case of a sudden storm and power cut. Use a long handled torch that will tuck under an arm. Used this way, the torch will give you a light wherever you need it. A lamp hanging from a hook, however well placed, always seems to cast a shadow where you want it least. It is dangerous to put a light on the floor during milking, for example, as it can be kicked over.

Mice will like to take up residence where there is plenty of food so discourage them by having a concrete floor that is easy to sweep.

The Exercise Yard
An exercise yard will be invaluable. When mucking out, the goats can be put there, thus enabling you to work without their assistance. If they get the opportunity, kids especially like to jump on the barrow.

Even a steep piece of ground can be used as a yard. The goats enjoy the variety of the view when it is terraced into uneven height steps. Always start concreting the lowest step first. Make provision for a food rack and a supply of water in the yard, as well as access to some shelter against the rain or very hot sun.

Mucking Out
Mucking out is a job of which one is aware all the time. It is always there, but provided it is attended to regularly – daily, weekly or annually – and provided you remember how it went in, in layers, it can be removed without any fuss or strained muscles. Get a muck fork as soon as possible. Why not suggest you are given one as a birthday or Christmas present? It may not be exciting or romantic but it is something you will use with gratitude almost every day. It has a long handle, the tines are dished and it is made to do this specific piece of work.

A well-made muck heap is worth its weight in gold, and does not smell. In order to keep it working (heating) gently, and avoid overheating, add muck little and often. Once you know the amount of muck to be removed from the goat house each time, you will be able to estimate the size of heap to make. Keep the heap neat and nothing will be wasted. If you add up to one barrow load each time, the heap will need to be about 3 feet (1m) square when the muck is spread thinly. Goat manure is high in nitrogen so it is worth taking the trouble to produce a good compost. You should eventually have three heaps: one being built, one completed and sealed with a 3-inch (7.5cm) layer of earth, and one being used on the garden.

A deep-litter system for the goats works very well provided they have access to a yard for some part of each day to cool their feet. Provided you can keep the doorways clear, this system is very little work, gives a good, clean, dry, warm bed and, apart from the very top layer, the manure can be put straight on to the garden or paddock in the autumn each year. The top layer will be too hot for this for up to a couple of months but can go on the compost heap as activator.

Hoof Trimming

Hoof trimming is another thing that has to be done regularly, and here again you must know your goat. Some hoofs need attention each week, but others can be left for a month. Neglecting to check hoofs can lead to foot-rot. A pair of secateurs, with a straight blade and anvil, is the safest tool for a beginner to use. Let the goat know what you are going to do by gently running your free hand down her leg, perhaps even giving her lower joint a couple of slight taps with a finger, and then lift her foot and turn it up so that the bottom of the hoof is clearly visible. The aim is to trim away the frilly edges and expose the flat sole. To do this, clear away any mud or muck, snip off the pointed tip of one half of her cloven hoof, thus enabling you to

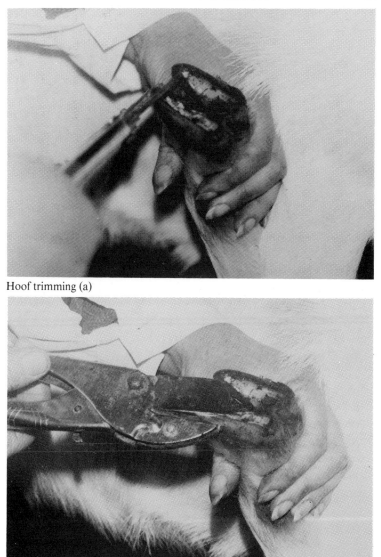

Hoof trimming (a)

Hoof trimming (b)

gently slip the point of the blade of the secateurs under the overlap of the horn flap. Keep the blade flat on the sole of the hoof and the anvil of the secateurs on the outside of the hoof and no harm will come to the goat or you. When one side of each part of the hoof has been trimmed, turn over the secateurs and trim the other side.

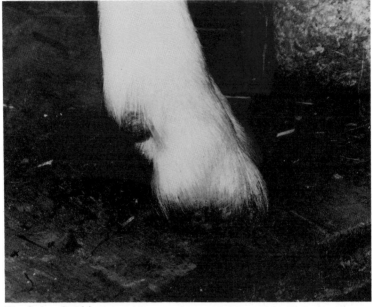

Hoof trimming (c)

Occasionally the heel needs to be cut back too. A knife or file could slip and you, the goat or both receive a nasty cut. In the unlikely event of the goat being cut, do not panic. Dip her hoof in some sterilizing water, and make sure the goat is put back on a clean dry bed until the bleeding has stopped. Stand the goat on a flat surface after trimming to check her feet are level.

General Tips
Whenever you are handling animals, make your movements

smooth and quiet. Train all children to follow your example. Only feed the goats at specific times. If titbits are given every time you pass the goats, they will come to expect them and become a perfect nuisance, worrying until you do give them something. When the goats crowd around a doorway, preventing it being opened, put your fingers under the chin of the nearest goat and gently raise it. She will have to go back if her chin is raised high enough and you will be able to open the door.

From time to time one of the goats is going to slip through a gate or hole in a hedge. As there is not usually time on these occasions to go and fetch a collar and lead, keep a piece of baling twine in your pocket or on a strategically placed hook so that no time is lost in bringing her home. Get ahead of her and this will tend to make her stop and turn round. If you chase her it will turn into a game which she will undoubtedly win! Once you have turned your goat towards home, she will walk more slowly. As soon as you are able to do so catch hold of and lift a back leg. Hang on to this until you have managed to slip your makeshift lead over her neck. Do not let go of her leg until you are sure you are holding her lead firmly. If the loop around her neck is loose, twist it once or twice until you can just get your fingers between it and her neck. It will suffice as a collar until you get her home and she will not have expended any energy running away from you, so that you will get the usual quantity of milk next milking. Your temper will be in a healthier state, too.

Make all tools and equipment do more than one job. Your secateurs do hoof-trimming as well as hedgecutting. Your barrow can be used for muck shifting, mixing concrete, carrying bags of food and hay: three bales of hay can be carried on a barrow, thus saving time and energy. A large curtain, well past its prime, is useful for carrying nettles and brambles when rolled up. Position your scales for weighing the milk so that you can also weigh the food.

3

FOOD AND FEEDING

Before going into this subject in detail, it is important to realize exactly why food is needed. Production, of course, comes immediately to mind, but what about heat and energy? Production is easy to gauge because this can be seen, but it is not so easy to estimate the amount of energy a kid is using when it is chasing around the paddock or yard in a scatty game. And production covers more than just milk. The growth of fur and flesh in young stock and renewal of coat and old tissues in the adults also has to be taken into account.

Fortunately, goats like a variety of food and this makes the work simpler. When listening to experienced goatkeepers discussing feeding, words such as concentrates, yield, roughage, starch equivalent, protein, carbohydrates, minerals, metabolism, get bandied about and leave the beginner more confused than ever. Obviously the goats have to be fed throughout the year but as the seasons gradually wax and wane the day-to-day routine changes very little, almost without our noticing it, in fact, so there is no need to panic about this.

Start by trying to identify the plants, shrubs and trees in the hedgerow. It is important to do this as some are better for the goats than others. Some are poisonous. Make a mental note of where each may be found. This saves time when gathering food during the winter. Bearing in mind that goats are browsers, not grazers, the free foods they prefer are hazel, ash, elm, sloe, willow, docks, brambles, thistles, ivy and holly (minus the berries), nettles, dandelions, chicory, and comfrey, to name but a few. Nettles make a completely balanced food, although some goats prefer them wilted to fresh.

Growing Goat Food

Gradually get a goat garden together. Start by sowing an extra row of carrots or a few extra brassicas. It will all help. Even weeds, once they have been identified, suddenly become valuable for goat food.

It is comforting to know that you are unlikely to kill a goat by overfeeding roughage, provided there is nothing poisonous in it. Roughage is the raw, fibrous, free food from the hedgerow and garden, plus hay. Unfortunately, you can kill by overfeeding concentrates or corn so always weigh each feed before giving it to the goat. The quantity of concentrates that it is safe to give her varies with the quantity of milk she produces, for example, she may have up to 5 lb (2kg) per day for every gallon (4.5l) of milk, $2\frac{1}{2}$ lb (1kg) per day if she only produces $\frac{1}{2}$ gallon (2.25l) and so on. These quantities should be divided into two feeds.

Many bought-in feeds contain 16 per cent protein, which is too high for goats. They only need 13 per cent protein, so if you are using these, particularly during the winter months, add crushed oats or flaked maize, which are low in protein but high in carbohydrate, until you get the feed to the correct proportions. It will help to remember, when mixing your own feed, that if a 16 per cent protein concentrate is used as a base, 1 lb (450g) crushed oats or flaked maize should be added to every 4 lb (2kg) of 16 per cent protein food to bring it to the correct proportion. Before dabbling with the mixture, however, always check the labels. There is a label on each bag listing the percentage of protein, fibre, minerals, and so on, but if you are worried about the accuracy of the feeds, ask the advice of a more experienced goatkeeper.

Learn the feeding value of as many foods as possible. A number of feeding value charts are available from agricultural organizations and in books. This information is worth accumulating, although some charts only list the protein and carbohydrate values whilst others include the mineral content. Charts will state whether the food is palatable, but very few say whether it is poisonous.

Goats can and do get bored with their diet, but whenever any change is being made, make it gradual. When the first feed of the change is being given, put in a tiny amount of the new food, so that they get little more than the smell.

This is particularly important when adding a mineral supplement. Introduce only a few grains to their concentrate ration and increase this to the full quantity over the next fortnight. All goats need a mineral lick within reach and this can be used as a guide to whether they need any extra. When you see them licking the block every time you pass by, a supplement is called for. Both chemical and vegetable based ones are available, and it depends on the time of year which one you use. Do read the feeding instructions. A goat that has periods of sucking her own teats is probably in need of a few extra minerals. Try giving her a course of seaweed meal. The self-sucking will probably stop within a few days. Putting a wide collar round her neck or mustard on her teats are only preventative measures. Extra minerals may cure the trouble.

Hay

Hay made from nettles is very much enjoyed during the winter, but hay is always an important part of the goat's diet. The most nourishing hay is that made during May, June and July. Allow 5 lb (2kg) per day per goat when estimating the purchase of the year's supply. The smaller the quantity purchased the more expensive it will be per bale. Hay bought 'off the field' is always cheaper than that from a merchant but you will need storage space. Some means of transporting the hay from the field also has to be found and this cost has to be added to the cost of the hay. Small quantities of hay may be made in a garden on a tripod. A tripod, made from 8 foot (2.6m) high poles, if loaded correctly, will hold up to 5 cwt (254kg) grass. The secret of making good tripodded hay is to keep a chimney of air running through the centre. If this gets filled the air cannot circulate and the hay will go mouldy. When the tripod is loaded, comb the top layer with your fingers to make it smooth. This will encourage the rain to run off and not penetrate the hay. Churchyards, old railway embankments, and orchards belonging to old-age pensioners are worth investigating for possible sources of unpolluted grass for hay.

Dusty or dank hay is bad for goats, so rather than feeding them with this, give them oat straw. Smell the hay before feeding. It should smell sweet. Feel it, too. It should be cool. Leave freshly made hay at least a month, and preferably six weeks, before using it.

Economic goatkeeping takes time and effort on the part of the goatkeeper; is this time and energy available or is the idea of goatkeeping only a dream? One should aim to give the goats some fresh food every day, regardless of the weather. The garden is a great help, and so are the neighbours, the local greengrocer, and possibly the local school. These folk do like their offerings collected regularly and often prefer you to provide your own containers. Always keep an eye open for knives and other implements that could have slipped in with the cabbage leaves or apple peelings. Surplus fruit is much enjoyed by the goats but only feed in moderation – for example, limit windfall apples to 12 oz (350g) per day.

Winter Feeding

Winter feeding is the most difficult, and it is therefore unwise to keep more goats than can comfortably be fed during this time.

Roots are useful during the winter months and these can be chopped to the correct 'chip' size by a specially designed root cutter, a clean sharp spade or a butcher's cleaver attached to a chopping block. Using the carving knife soon becomes tiring when feeding more than a couple of goats. Many goats like their roots sprinkled with a little broad bran to dry them off.

A hand operated root cutter

Another winter feed they enjoy and which will help to keep up the milk supply is sugar beet pulp. This should be fed with care and caution. It must always be remembered that sugar beet pulp swells when it comes in contact with a liquid. As a safeguard, soak the pulp in hot water before use, let it absorb all the water it will and then drain and squeeze out any surplus. This surplus water may be added to the goat's drinking water. When the pulp is cool enough, it may be fed to the stock. Sugar beet pulp should only be fed dry if the goats are individually rationed for it and allowance is made for swelling when it comes in contact with the stomach juices.

Before feeding kale or any of the brassicas, give them a good shake to remove any rain or dewdrops. Brassicas fed wet can make the animal scour. When this happens, at the first signs of soft or runny droppings, put the patient on to a diet of nothing but hay and warm water for twenty-four hours, even if it means shutting her indoors for this period. Let her have as much hay and water as she wants, but nothing else. Check that she can reach her mineral lick, though.

One way to bring home the kale

Industrial wastes can make a considerable saving on the feeding costs. Grass nuts, 16 per cent protein, are often available from a contractor at the local airport; milk factories sometimes have tins of rice pudding, 3 per cent protein, available for animal consumption; apple pumice, a carbohydrate, and brewers' grains, 20 per cent protein, can often be found locally. The latter two items should be stored in plastic bins and be well compressed, say with a heavy piece of wood, and pummelled down between layers, and they will stay sweet for several months. Each area will yield its own particular industrial waste. Search it out.

Another way to economize is to buy in bulk, with two or three goatkeepers pooling their order. Collecting the order when passing, instead of having it delivered or making a special journey, is also sensible.

The majority of goats in the UK will have to be yarded twenty-four hours a day during the winter because of the weather, but do let them have access to the fresh air so that they get their quota of vitamin D.

Spring Feeding

When the spring comes, if the goats have had very limited browsing during the winter, only introduce them to the lush grass for a brief period of about fifteen minutes. After three or four days, increase it by five minutes. It is safe to increase this period by a further five minutes each day from then onwards, but as a further safeguard make it a strict rule that the goats always have some hay before they go out, to prevent them gorging.

Gorging lush or wet grass can make them hoven, or blown, which is caused by an excess of stomach gasses not being able to get away quickly enough. This makes the stomach very distended so that it feels like a tightly stretched drum to the touch. It is obvious that the animal is very uncomfortable. Drench the patient with $\frac{1}{2}$-1 cupful of liquid paraffin (mineral oil), preferably slightly warmed. If this is not readily to hand, use baby's Gripe Water or a proprietary brand of stomach powder. If none of these things is available, fizzy lemonade will suffice. Taking her for a walk will also help, but do not let her stop to nibble.

Drenching

Practise drenching a goat with warm water well before an emergency arrives. This will give you confidence. Use a container that will nestle comfortably in the hand. A container, say a small teapot, with a lid and a handle through which you can slip your thumb to anchor it, is best. If you are right-handed, cup the container in that hand and putting the goat's neck between your body and your left arm, gently place your left hand under her chin. With your left thumb, pull down the side of her lower lip so that you can just insert the end of the spout or lip of the container, and slowly tip the medicine alongside her jaw. Should you be over-enthusiastic or too impatient and the goat starts to choke, let her head drop immediately so that she

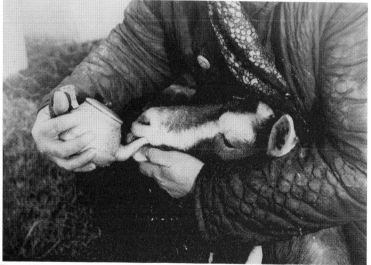

Drenching

can get her head down to cough. Otherwise the liquid could get into her lungs. Alternatively, use a large syringe minus the needle. Your vet will probably let you have one.

Drinking Water

Milk contains 86 per cent water so particular attention should be paid to what she drinks. The quantity she will need is approximately $\frac{1}{2}$ gallon (2.25l) per day plus four times the quantity of milk produced, but should her supply become fouled she will drink none of it. Water that is slightly warm always encourages her to drink more. During the winter, warm the water midday as well as morning and evening. Some kids will drink water when it is really hot but common sense must be the guide at all times.

Food and water containers must be kept clean. Make regular washing a routine procedure. Water buckets very quickly get an almost invisible slime on the bottom if one is not careful.

Using Your Land

Where there is very little land to spare for the goats, it is wisest to crop it. If it is grazed without a break it will become infested with worms. Small quantities of farm seeds are available if you shop around for a supplier. Kale, swede, fodder beet, mangold (toxic to the male only), tick or horse beans (plant as broad beans), maize (to be fed at the tassle stage), are just a few things that can be grown for the goats. By first using the thinnings of kale for the goats, and then the leaves when they are large enough, the toppings next, and, finally the remainder of the stalk, this crop will be spread over several months. Nettles, too, will provide four crops of useful food a year if they are harvested carefully.

Very few goats have the opportunity to browse in woodland or over heath or moorland, but access to a herbal ley will compensate them for this. A herbal ley contains more than just grass and a dash of clover. It will also contain many deep-rooted plants. These are greatly appreciated, especially during drought. A herbal ley also contains a number of fibrous plants which are of great benefit to the goats. Where it is not possible to convert an existing paddock to a herbal ley, small patches, edges of driveways or a strip of the kitchen garden can be used and it can then be fed to the goats in small quantities when it is over 9 inches (22cm) tall. Do not feed it to the goats when it is wet. Let it dry a while, as you would for grass.

Many people who start their goatkeeping in a very small way, with a yarded system in a town, do eventually decide to look for an acre or two of land in the country. If you decide to move your goats to a new area, walk over the land, and garden, to identify and, if necessary, remove and burn any poisonous plants before allowing the goats out of their new house. Many things in the flower garden are poisonous so a few repairs to the fence around this could be advantageous. Flowering shrubs and evergreens, with the exception of ivy and holly (minus the berries) will head the list.

Rhododendron poisoning is believed to be the only thing that makes the animals vomit, but as this is not their usual method of elimination it does seem to cause a considerable amount of pain. Rhubarb, beetroot leaves, daffodils, ground elder, tomato and potato haulms, laburnam and sycamore, as well as ragwort, hemlock and foxglove can also be disastrous. Some poisonous things can kill very quickly, so don't delay in contacting the vet if you suspect poisoning. The best solution is to be alert in the first place. However, once the goats are used to ranging free, especially the offspring of the original arrivals, they will learn to find many antidotes themselves. Acorns are an example of this. Acorns on their own are poisonous but where the goats have access to grass the quantity of the milk yield will be depressed but nothing more disastrous will result.

Metrication

Now that we are becoming ever more closely linked with the other European countries more things are 'going metric'. Goat foods and feeding values are no exception. In future food will be evaluated in megajoules (MJ) per kilogram (kg) and not starch equivalent and protein. Feeding charts and labels will increasingly show the value of metabolizable energy (ME) instead of those with which we are all familiar. At first this method of evaluation might seem complicated but start to look for information on this subject now so that by the time it is common practice you will be fully prepared and able to cope.

4

MILK AND MILKING

So often we hear that 'Goat's milk is good for you', which is true
but such a statement needs qualifying. The main differences
between goat's and cow's milk are:

1. The fat globules in goat's milk are smaller than those in
   cow's milk, which makes it easier to digest. Young babies
   and old folk especially appreciate this.

2. Because of the smaller fat globules the milk may be deep-
   frozen satisfactorily, unlike cow's milk which tends to
   separate when it is removed from the freezer. Milk for the
   freezer must be cooled quickly (within twenty minutes)
   before it is put in there. This ability to deep-freeze the milk
   has many advantages, not least during times of surplus and,
   of course, correspondingly, prior to kidding when fresh milk
   could be in short supply. It also means that customers who
   also own a deep-freezer, are able to collect bulk supplies of
   milk instead of calling every day for it.

3. There is less lactose (sugar) in goat's milk than in cow's
   milk, and diabetics, or anyone on a diet of restricted sugar
   intake, may drink 4 oz. (100g) of goat's milk for every $3\frac{1}{2}$ oz.
   (87g) of cow's milk prescribed.

4. Goat's milk does not contain the protein which, in cow's
   milk, is thought to be one of the causes of eczema and
   psoriasis. Certainly, patients suffering from these

complaints show a marked improvement when goat's milk is substituted for cow's milk and other bovine products are removed from their diet.

5. The calcium in goat's milk is more easily assimilated into the human body than that in cow's milk.

6. There has been no recorded case of brucellosis in goats in the UK for over fifty years.

7. Goat's milk is the nearest commonly available milk to human milk.

Old wives' tales abound on the subject of goats and it must be admitted that many do have a grain of truth in them. However, many originated through ignorance and the others have been exaggerated beyond recognition because they made a good story. It is, therefore, up to every goatkeeper and potential goatkeeper to see that nothing they do encourages the spread of these tales, many of which concern goat's milk.

Imagination plays a very big part in people's minds when they first taste this milk. If they have not been told that it is goat's milk, and if it has been produced under hygienic conditions, backed by proper feeding and management, very few will pass any comment at all, beyond saying 'That was a nice cup of tea'.

Correct Preparation

It is not difficult, or particularly expensive, to achieve palatable milk but it does take a little time and care each day. Every utensil connected with the milk and milking must be sterilized twice each day during the summer and at least once during the colder weather. But as it would all have had to have been washed twice each day anyway, this means no more effort than adding the appropriate sterilizing/detergent agent to the water instead of washing-up liquid and rinsing everything in cold water afterwards.

Most clubs will be able to supply small quantities of a sterilizing/detergent powder but, if they cannot do this, perhaps two or three goatkeepers could purchase a small drum and share it. The majority of powdered sterilizing agents do have a limited shelf life of about a year, hence the need to share a drum, but they seem to be the most popular because of the ease of handling. Read the instructions on the label carefully before using it for the first time. When the sterilizing has been completed, any mix that is not being set aside for emergency use may be used for swilling down the milking bench, floors and exercise yard, but do avoid leaving puddles where an iodine-based sterilizing agent has been used. These are first class for the job for which they were designed, but they can be lethal if chickens and other small livestock take a drink.

Where time does not permit sterilizing to take place immediately after milking, rinse everything in cold water. If this is not done it is much more difficult to sterilize properly later on.

The actual milking may be done by hand or machine. Hand milking is a pleasant occupation and until sheer numbers drive one to the machine most people prefer this method. Whichever method is used, the question of washing the goat must be decided on. If washing is desired use running water, say from a boiler, as hot as your hands will allow, wring out the cloth and wipe just the teats gently but thoroughly, and dry the area with a soft, small towel. For the second and subsequent goats hold the udder cloth under the running water again and wring out. Do not put the cloth in the water bucket and mix up the 'foreign bodies' there. You could be transferring on to the next udder as many as you have taken off. Unfortunately, not all foreign bodies can be seen. Before starting to milk make sure, too, that there are no pieces of hay on her tummy to fall into the milking bucket. As a precaution, pass the back of your hand under her. Finally, if your own hands are spotless you are ready to begin milking.

If you are comfortable and relaxed this will be communicated to the goat. Sit as nearly parallel to her as you can, with your right shoulder almost under her chest — this is if you are right-handed, and that is the side that is most convenient for you. Sit the other side if you are left-handed, of course. You should remember, though, that goats are not keen to be milked from the opposite side to that from which they were milked originally. Check this point when buying an adult milch goat or life could prove difficult for a while.

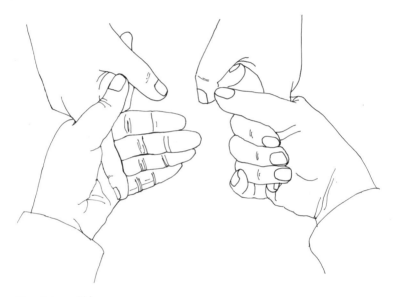

Practising milking

The Technique of Milking

You should now be ready to begin milking. Gently, but firmly, place your hands around the upper half of the udder and then slide one hand down each side until a hand is on each teat, all the while consciously feeling for any lumps, bumps or scratches which must have attention. With your thumb, rub away the small particle of dried milk that will have sealed the bottom of the teat since the last milking twelve hours ago. Put the bucket into position and, if the teat is big enough, seal the top of one teat, just below the udder but not on it, with your thumb by trying to press it flat. Keep it sealed while your index, middle, third and little fingers are flapped, pressed and held across the teat. Use your middle finger to seal the top of the teat if it is a small teat, but if the seal is not held all the way down as the other fingers come across, you will feel the milk trying to go *up* hill. If this happens too often the udder may be damaged. (When learning to milk, it helps to think of a paste coming out of a tube.)

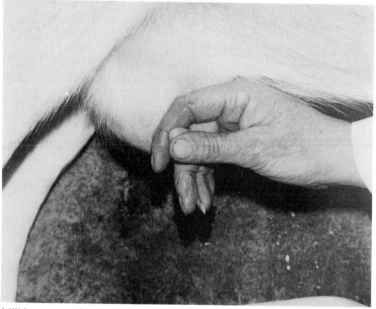

Milking – removing the teat seal

When one jet has been extracted, release your hand from around that teat and take a jet from the other teat with your other hand. Repeat the action with the first teat, then the second, and so on. A great deal of concentration is needed at first but

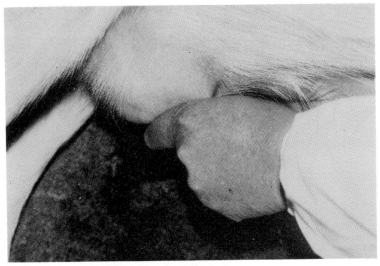

Milking – correct position of hands on the teats

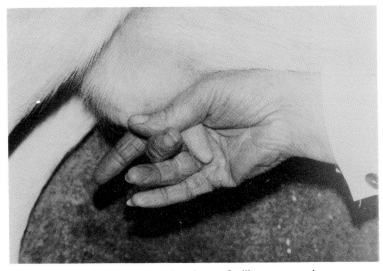

Milking – 'stripping' to ensure the last drops of milk are removed

gradually you will notice that a rhythm is beginning to come. Singing a marching song is a help too; one such as *Onward, Christian Soldiers* is good for getting a rhythm going.

Carry on in this way until the lower section of her udder and teats look like an old glove and no more milk can be extracted. No pulling, banging or slapping is necessary to achieve this, just non-stop, steady milking. Occasionally one quarter will need a few extra flaps, but if any milk is left behind your goat will think it is not wanted and begin to go dry. Milking is repeated at as near twelve-hour intervals as one can possibly make it. It does not matter if the milking takes place at 6 a.m. and 6 p.m. or midday and midnight so long as there are twelve hours in between.

In the beginning it will take some time to milk her – perhaps half an hour – but with practice it will take as little as two minutes for every pint of milk she gives you. Practise milking on someone's index fingers. They will be able to feel whether or not you are maintaining a seal and steady pressure. Say to yourself 'Seal, flap, release; seal, flap, release ...' Confidence, through knowledge, is a great help.

Milk Production

As a very simple illustration of the way milk is produced, compare this with a cook making a cake:

The goat eats her food and this goes into the first of her four stomachs where it is broken down by the formation of amino acids and eventually enters the blood stream.

We collect the ingredients for a cake, put them in a bowl and mix them with egg and milk to make a dough.

When your goat learns your routine she will know when to let-down her milk.

When the dough reaches the correct consistency we put it in a tin and so into the oven.

Milk is not stored in her udder, although it is collected in the cistern, the lower section of her udder, during let-down, prior to milking.

That which enters the udder is not milk. It is only milk when it has entered and passed through the udder.

When the dough is put into the hot oven it is not cake. It is only cake when it is taken out.

The udder is like a factory which is full of machines called lobes. These come into production as and when required to do so by the stimulation of the birth of the kid. They will continue producing milk whilst the need is there. Hence the same udder looks larger in the summer than in the winter. Goats that run-through for a second year do so because they have been encouraged to do so by breeding, feeding and milking.

Keeping a Milk Yield Record
Immediately after milking weigh the milk and make a note of it, as the goat's concentrate ration depends on this. One of the simplest ways of recording is on a graph. Once it is set up with dates and weights it will only involve the placing of a dot, spot or marker in the morning and a short dash in the evening linking the total of one day's milk yield to the next. The a.m. marker just 'holds' the yield until the p.m. milk is added to it.

It will be noticed from this that after kidding, provided the kids are not left on the goat, the yield will gradually rise for at least the next six to ten weeks. It will then level out and only gradually decline in the autumn as the feeding value of the fresh roughage declines, level out again during the winter but start to rise again the following spring and summer as she has more and more access to fresh food. During the second spring and summer of each lactation her yield will not be as great as the previous year but should reach two-thirds of it.

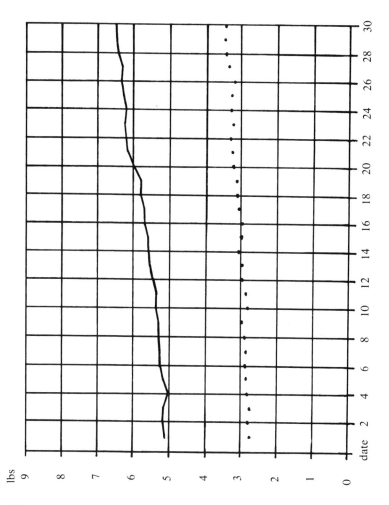

Milk recording graph

Straining and Cooling the Milk

After weighing the milk should be strained. Use a proprietary brand of filter, readily available from most agricultural merchants, for this. Butter muslin will only extract some of the hairs and hay and is not easy to sterilize properly, whilst the sterile paper pads can be thrown away after each milking and there is nothing to worry about at all. However, should a foreign body fall into the bucket, or whatever container you are using for milking, *leave it there*. As the milking progresses a froth will form on the top of the milk and bring up the foreign body with it. When the milk is poured through the filter, whether this is just a filter paper in a kitchen sieve resting on a large jug, or a filter paper between two metal straining discs in a plastic funnel resting on the rim of a carrying can, the foreign body will stay with the froth, probably around the edge of the milking container and no harm will be done.

If you try to extract it whilst milking it will probably be broken into more than one piece. You are then left with the even more difficult task of extracting at least two small objects instead of one large one. When the foreign body does fall in, be very careful that the next jet of milk does not push it through and under the froth.

The milk must next be cooled. Standing the container in a running stream or well is a romantic idea but unfortunately, whilst this does work very well, not everyone has such things to hand. A refrigerator does not do the work quickly enough. During the hot weather it is sometimes necessary to cool a bowl of water in the fridge and stand the fresh milk in that. At other times tap water is sufficiently cool, but check the temperature from time to time to be sure. Aim to get the milk down to a temperature of 50°F (10°C) within twenty minutes of milking.

An alternative method of cooling that is very efficient is to connect a sparge ring to a cold water tap with a piece of plastic tubing and place the ring over the outside of the carrying can.

The water then trickles through the holes in the ring and over the outside of the whole of the can, keeping it really cool. A sparge ring may be made from a piece of plastic tubing that has had holes drilled in it at 1-inch intervals on the underside. The tubing is joined into a circle with a 'T' junction, obtainable from any plumber. The third end of the 'T' is used to link the ring to the tap with the appropriate length of flexible pipe.

Finding the Reason for Tainted Milk

If the foregoing routine is followed, the milk should be palatable 365 days of every year, but if something goes wrong and you find yourself with gallons and gallons of milk that nobody will drink because it is tainted, action must be prompt and efficient. Only one pint of unpleasant milk has to be sold to do harm to every other goatkeeper.

Inspect every utensil that is used in connection with the production of milk and check on storage and sterilizing habits. If these are correct, questions will have to be asked about the feeding and health of the animals. Can the taint be identified? Is it bitter? Is it goaty? Does the milk smell? Is it from one goat or all of them? Do the goats need worming? Do any of the goats have unpleasant breath? What were they fed during the four hours prior to milking? Get some of these questions answered and the problem is halfway to being solved.

1. To find out whether it is one or all of the goats that is causing the trouble, milk the first few jets of milk into a small receptacle and sip it while it is still warm. The taint will be emphasized then and easier to identify.

2. If it is all of the goats that are producing unpleasant milk, suspect the feeding. Have they been fed kale, which makes the milk taste of cabbage water, within four hours of milking? If so, make sure everyone understands that it is better to feed this only *after* milking. Hazel can give a nutty flavour to the milk and whilst most people enjoy this, not

everyone does. Neither of these taints are harmful, they are just not to everyone's taste.

3. If it is absolutely certain nothing is wrong with the sterilizing practice but the milk tastes goaty, the fault could be that it has not been cooled quickly enough. Check the length of time this has been taking.

4. A deficiency in the mineral or vitamin content of the goat's food could be the cause of other taints. Lack of vitamin B12 or cobalt both cause a bitter taste in the milk. The first can be dealt with by giving her a sandwich containing a yeast spread, available from grocers and supermarkets. The latter needs a solution of cobalt crystals, obtainable from the majority of dispensing chemists, poured on to a salt lick over a period of ten days.

Anything less common than these taints will probably have to be investigated by the Ministry of Agriculture Veterinary Department, although a goat whose milk yield is dropping for no apparent reason, has a starring coat, always seems to be ravenous, and whose breath smells does need worming and could produce milk that is not 100 per cent palatable.

A drop in milk yield frequently occurs when a goat is in season but this is nothing to worry about. On the other hand, if it occurs during the first six weeks after a kid has been separated from its dam, check all door and gate catches. The kid could have managed to get back for a quick snack!

5

GOAT'S MILK PRODUCTS

Once the kids have been born, and the goats can get out to browse for the best part of the day, there will be surplus milk available for cheese. In the U.K. cheese should be made during the months of June to September. During the colder months the cheese will not drain properly unless artificial conditions are created.

Goat cheese is better if a starter is used. This is made by taking a freshly sterilized screw-top jar, milking some milk straight from the goat into it, and sealing the jar immediately. The milk must not be cooled but should be left in a warm place — say, on the back of a stove — for a couple of days or until such time as it changes in appearance. It is then ready for use.

Soft, semi-hard and hard cheese may all be made under home conditions. The soft cheese is often referred to as cottage cheese and up to 2 lb (900g) may be made from each gallon (4.5l) of milk. Half this quantity of coulommier, the semi-hard cheese, may be made from a gallon of milk but it will take about 3 gallons (13.5l) of milk to produce a 2 lb (900g) hard cheese. Cottage cheese can be stored in the deep-freezer but the coulommier should be eaten fresh, or within a week of the process being started. If necessary, it can be stored for a few days longer in the refrigerator. With plenty of cool air circulating around it, a hard cheese will stay fresh for a considerable time.

Cottage cheese does not need rennet, but the other types do. Only drips are added to each gallon of milk so there is no advantage in purchasing a vast quantity. It does have a limited life even when stored in a cool, dark place. A vegetable-based renetting medium is now available.

Cottage Cheese

Add a little starter to the milk available for cheese and place all the milk in a container which can be left in a warm place for two or three days. Keep this covered until the milk has curded. With a fine meshed material (a clean, old nylon net curtain will do), strain the curd and hang up to finish dripping. When the cheese is damp, but not wet, it is ready for use or storage. Salt may be

Cottage cheese – curd ready for draining

added if desired. Alternatively, chopped chives or grated nuts add interest and flavour.

The whey may be used to feed the chickens or pigs.

Coulommier Cheese

A thermometer, suitable for making sweets or bread, will have to be obtained before coulommier cheese can be made, but all the other equipment can either be located around the home or bought for a few pence. Moulds can be made from family-size baked bean cans, if both ends are removed. The outside rings of some cake tins are also suitable. Balsa wood and straw cheese mats are not often available nowadays, but pattern/colour sample pieces of plastic kitchen floor mats make good substitutes and

Coulommier cheese

may be obtained from ironmongers or furniture stores.

Heat 1 gallon (4.5l) of milk to 84°F (29°C) preferably in a water bath, and add 2 tablespoonsful of starter. Mix 1 teaspoonful of *cheese*, not junket, rennet with 6 teaspoonsful of cold water. Add this to the milk and stir for 2 minutes. Curds will then form. Leave for 30-45 minutes to set. Scald all equipment – boards, mats, moulds. When the curd is firm, ladle it gently into the moulds and leave in a warm room to drain. Drainage will take approximately 2 days but each day turn the cheese, still in its mould, on to a fresh mat and board. When the cheese is firm, remove the mould. It can then be salted and is ready to eat.

A smaller quantity of milk may be used but do not forget to reduce the amount of starter, rennet and water pro rata, although temperature and times will stay the same.

Hard Cheese
A little more equipment is needed for this but one should be able to adapt or make what is necessary from what is to hand around the house.

Sterilize all equipment, heat 3 gallons (13.5l) of milk to 158°F (70°C) and cool immediately to 90°F (32°C). Add 5 fl. oz. (150ml) starter and stir thoroughly. Leave to ripen for 30 minutes. Add 4 teaspoonsful tepid water to 1 teaspoonful rennet in a sterile container and mix together. Pour this into the warm milk and stir thoroughly but do not overstir. The curd will form and be firm in about 40 minutes. Using a sharp, double-edged knife cut the curd into $\frac{1}{2}$-inch (1cm) wide strips. Then, cut it at right angles to form $\frac{1}{2}$-inch (1cm) pillars of curd. With a long, flexible knife, cut the curd horizontally. Finally, loosen any curd around the edge of the pan. Leave for 5-10 minutes until some whey has appeared on top.

Gently stir the curds and whey with your hand for half an hour, gradually heating the mixture to 100°F (38°C) while doing so. Do not let the mixture exceed this temperature. At the end of 30 minutes, remove from the heat, but continue to stir for a further 5 minutes. Allow the curd to settle to the bottom during the next 30 minutes.

Hard cheese

Ladle off most of the whey. Tip the rest of the curds and whey into a sterile cloth (sheeting will do if it is sterile) and make this into a bundle by winding one of the corners round the other three. Put the bundle on the tray which has been tilted at an angle to allow the whey to run off. Untie the bundle after 15 minutes, cut the curd into 4 long slices and pile these up. Retighten the cloth as before. Release, and retighten the cloth a further two or three times at 15-minute intervals after the curd has been cut and restacked. When the curd is quite firm break it into small pieces about the size of a nutmeg, sprinkle 1 oz. (25g) of salt over it and toss the curd.

Boil a square of muslin for 10 minutes, line the mould with it and stand this on a tray. Press the curd into the mould firmly, cover the top with a corner of the muslin and put the wooden follower on top. A follower can be made from a piece of wood 1½-2 inches (4-5cm) thick, which will fit snugly, but not too tightly, in the mould. Place a heavy object, about 28 lb (13kg) in weight, on top of the follower and leave overnight.

Next day, remove the cheese from the mould, dip it into hot water (approximately 150°F or 68°C) for 15-30 seconds, and wrap it in a fresh muslin square. Return it, upside down, to the mould. Increase the pressure on the follower to 56 lb (25kg).

On the third day turn the cheese again and return it to the mould, but by the evening it should be firm enough to take out of the mould.

This cheese will be ready to eat in three or four weeks' time. Allow a longer time if more flavour is desired. To prepare the cheese for storage cut a piece of muslin as wide as the depth of the cheese and 1½ times the circumference in length and 4 pieces of muslin to act as caps (2 each end). The caps are stuck on each end of the cheese with lard, or other edible paste, first of all and then the 'bandage' is wrapped firmly around the outside and stuck down as it goes round. Aim for a storage temperature of 51°-59°F (11°-15°C). Turn daily.

Yogurt

This has become increasingly popular and is simple to make at home. A ¼ pt (150ml) container of plain yogurt will act as the culture for eight further ¼-pints.

The method is to boil 2 pt (1 litre) of milk in a double saucepan and to keep it simmering for 30 minutes. Cool the milk to exactly 113°F (45°C). At this stage pour the milk on to the culture and seal the container(s). A thermos flask is good for this. Alternatively, use a baby's bottle warmer or stand the container in the bottom of the airing cupboard overnight. The mixture should be left in the container for at least six hours, but the actual length of time will depend on your own particular taste preference.

Further batches may be made using $\frac{1}{4}$ pt (150ml) of your own yogurt for the culture although in time this will get progressively less acid and a fresh supply will have to be bought.

Butter

Goat milk butter is delicious but due to the small fat globules in the milk it takes a long time for the cream to rise to the surface. It stays in suspension. Therefore, unless they have a cream separator many goatkeepers prefer to convert their surplus milk to cheese. Those that do make butter have no difficulty in storing it though. It is eaten so quickly.

Cream separator and 1 gallon butter churn

Stir in each day's cream to that collected previously until a sufficient quantity has been accumulated and keep in a cool place (56°-60°F or 13°-15°C). Add no more cream for the twelve hours prior to churning. To prevent sticking, stand the churn in cold brine made from 1 lb (450g) salt to 1 gallon (4.5l) of water. If desired, colouring may be stirred into the cream before it is churned.

Add cold water to the cream until it forms a thin 'batter'. Fill the churn no more than one-third full with this and start to turn the handle. When the cream thickens add some more water. Repeat the churning and cold water process until the cream 'breaks' – separates into butter grains and buttermilk.

The butter now has to be washed. Pour cold water into the churn, turn the handle once or twice only and pour off the liquid. Do this as many times as is necessary to produce clear water and till there is no buttermilk left. Finally, add $\frac{1}{2}$ gallon (2.25l) water containing 8 oz (225g) salt and leave for about ten minutes. Drain off as much water as possible then.

To remove the last of the water, pat the butter with what are known as Scotch Hands. Only experience can tell you when to stop but if the water is not removed the butter will not stay fresh so long. If it is bashed too much, on the other hand, the butter loses its texture.

If a churn is not readily available, an electric mixer geared to the slow speed may be used for small quantities. The boxes of oddments at sales of small family farms will often produce butter-making equipment.

Marketing

Marketing dairy products should only be undertaken after considerable thought has gone into the subject. Not only must you be able to guarantee high quality products, but you must also be able to guarantee a supply of milk 365 days of the year. A supply of deep-frozen summer surplus may well be enough for the goatkeeper and his family, leaving the freshly frozen milk for the customer.

During long journeys in heated vehicles, fresh milk can deteriorate quickly due to the great fluctuations in temperature. Where it is suspected that the milk will have to travel in this way, suggest that it is carried in a specially designed box, perhaps containing an ice pack, such as is used for carrying ice cream. Now that they are comparatively inexpensive, many families do possess one.

The presentation and packaging of each product is crucial to its sale. Waxed cartons for milk are obtainable in small quantities from many clubs, and larger quantities (20,000) may be obtained direct from the manufacturers. There are now suitable plastic bags available but you will need to buy a heat sealer. Always resist the temptation to use bottles, jugs or other containers presented to one on the doorstep for filling. It is most unlikely that these will be sterile. Many squash bottles retain a smell of their previous contents for some considerable time.

The rules and regulations governing the retail sale of goat's milk are very few but a stamped measure is essential. Stay alert for changes in the law at all times and check with the local authorities for any by-laws, and with the Weights and Measures Department and the current Food and Drugs Act for anything concerning the sale of dairy goat products.

Should the retail outlet be a shop of any description, the local authority may contact you to say they have a sample of your product and are in the process of testing it. Welcome this service. The results are usually indicated to you within fourteen days. The Ministry of Agricultural Advisory Service will also undertake milk tests if you wish. The advice is given free of charge but a charge is made for all laboratory tests.

6

KIDDING AND WHAT GOES BEFORE

A milker's milk yield continues to decline after mating and she should be dry at least six weeks, and preferably eight weeks, prior to kidding. If she needs help for this, start to leave a drip or two of milk behind when you milk her. Being an intelligent animal she soon realizes what is happening and starts to produce less and less milk. *Never* be tempted to starve her dry. Milking at slightly lengthening intervals often helps if the previous method fails.

A goatling, a first kidder between one and two years old, should only be receiving 8 oz (225g) concentrate ration per day when mated. Six weeks later, when you are as sure as you can be that she is in kid, gradually increase her ration so that she is receiving the quantity needed to supply the milk she is capable of producing at her peak by a fortnight prior to kidding – that is, if your investigations suggest she will produce 1 gallon (4.5l) of milk at her peak she needs to receive between $3\frac{1}{2}$-4 lb (1.5-1.8kg) concentrates per day, split into two feeds, four and a half months after mating. Working from 8 oz (225g) per day, an increase of 4 oz (100g) per week is about the ideal. Be accurate with your weighing and increase the concentrates regularly – say, every Sunday or every Monday – so that you avoid mistakes.

A milker must of course be fed according to her yield in the early days of her pregnancy, but after the halfway stage feed her to her potential yield, whichever is the greater. Any increase in her ration must still be made gradually, however.

Physical Changes

As kidding approaches, the milker will definitely begin to look more rotund but a goatling may not show any signs at all for at least three months, and possibly three and a half months, after mating. During the latter few weeks of her pregnancy it will be noticed that her flanks look thinner and her tail gets higher. Just prior to kidding you should be able to close your thumb and index finger around her backbone, about an inch (2.5cm) up from her tail. These are nature's ways of preparing for the birth. Some goatlings' udders start to develop fairly soon after mating but others do not until the last few weeks of the pregnancy. During an older goat's rest period her udder is comparatively small, but gets much bigger the nearer she gets to kidding. If her udder does get hard and shiny and she is obviously uncomfortable, milk out a little colostrum but no more than $\frac{1}{4}$ pt (150ml).

If you wish her to kid in a different place or pen from her usual bed, change her to her new place three of four weeks before kidding if possible so that she is well and truly used to it. Keep everything as calm as possible before and after kidding. Never leave buckets of water in her pen from ten days before to at least ten days after kidding unless you are looking for a drowned kid. Just offer her warm water at regular intervals throughout the day. She will soon get used to this slight change of routine.

Most goats kid easily and on the due date but if the weather is cold, snowy, stormy or just plain horrid, the dam may not drop her kids for another three or four days. On the other hand, if it is beautiful, and likely to stay that way, she may kid three or four days early so be prepared for either eventuality. Listen to the weather forecast.

Your goat may get a bit restless as kidding approaches and scrape up her bed, sit down and then get up quite quickly. At this state you will probably notice her udder has flooded, or suddenly filled, and looks much bigger and heavier. She is about to kid. From the beginning of the kidding day until she has actually finished her work, visit her every half an hour and make sure you have a small towel or piece of clean, soft rag handy to wipe the kid's nose and mouth free from mucus.

The first drink

If you are lucky enough to be there when she kids you will see the two, tiny front feet appear first, then the nose resting on the legs and, after one or two more pushes, the whole head will come. Another push and the shoulders are out, and the rest of the kid just slips into the world. Most kids lay there for a split second, and then shake their head and sneeze. That's as it should be. You can then finish the job of wiping their faces but let Mum lick them dry. If there is a second kid it will arrive within ten to twenty minutes as a rule. The kids are often on their feet and sucking within half an hour. Never go to bed yourself until you are sure the kids have had a drink of colostrum.

Goatlings usually have only one kid and the majority of older goats have two but three, four or five kids are not unknown. But frequently one thinks a third kid is on the way when it is the afterbirth. Be particular in checking on the arrival of this. If it hasn't come away within twenty-four hours contact your vet. If only part has dropped never, ever try to pull out the remainder. Wait twenty-four hours and ask your vet's advice. Remove the afterbirth from the pen as soon as it drops. The contents are nutritious but the dam could choke herself if she eats it. A muck fork is the best thing to use for its removal.

After kidding, so as not to disturb the dam too much, just add some more dry bedding, make sure there is a rack full of good hay, offer her a drink and, if you have the will-power, go away and leave the family in peace for a couple of hours.

Difficult kiddings do occur and the assistance of another, more experienced goatkeeper will give you confidence, but take things slowly. Remember, only help if you must, only pull when she pushes and always pull in a circular movement towards the udder. Avoid going inside the goat if you can but, if you have to, scrub your hands and arms well, make sure your nails are very short, and lubricate your hand with plenty of soap before you enter. Think carefully and don't panic. There isn't much room so a lady's hand is better than a man's for this job.

7

KID REARING

Kid rearing begins from the moment the kids are born. It is essential that the kids are fed by the dam for the first four days of their life as the colostrum they then receive is full of antibodies to give protection from disease. It contains vitamin A and is also laxative.

The kid's first droppings are jet black but as it starts to drink the colostrum they turn golden/orange and will stay this colour until their diet changes, when they will eat the odd piece of hay or leaf that has dropped from the rack.

When the kids are first born the colostrum is yellow, thick and sticky but by the end of the fourth day it has changed to milk and is suitable for sale. Prior to this it is difficult to use in cooking as it curdles when the temperature approaches boiling point. Should it be obvious that the dam has a surplus, ease her by milking no more than $\frac{1}{4}$ pt (150ml) and store this in the deep-freezer in case of an emergency with another kid. Some shepherds will also be glad to know that you have a small supply available.

Apart from ensuring that the kid is sucking well and is quite lively during its waking moments, you should check that the dam is well too. After the afterbirth has been removed, continue to make sure the dam has plenty of roughage and watch to see that her udder is not overstocked and tight. Sometimes one side needs to have some milk eased out but take no more than will make her comfortable. *Never* strip her out completely during these first four days.

If the dam repeatedly pushes away the kid, try to find out why she is doing this. It could just be that the kid has sharp teeth and the teats are sore. Apply a little udder cream until the wound has healed. Anything more serious probably needs the vet's attention, but the vast majority of goats are good mothers.

By the fourth day many things will have had to be decided, not least the method of rearing. Is the kid to be left on the dam and the dam stripped out twice each day to stop her milk decreasing, or will the kid be bottle-fed? Both methods are acceptable. It is, therefore, advisable to adopt the one that suits your own routine.

If the kids are to be bottle-fed they need to be separated from the dam for six weeks or they will keep returning to her to suck. After six weeks there should be no more trouble. Put the kids where the dam is able to see them but not reach her udder. If they do get back, the six-week separation will have to start again! Kids reared this way are very friendly and one does, of course, have control of the feeding, whereas when they are left on the dam she is the centre of their existence and one can only assume they are getting adequate food by their appearance and behaviour.

If a Kid is to be Sold

The next question to ask is, 'Is the kid to stay with the herd or be sold?' If the latter, keep her until she is a month old at least. She will grow better if she is able to have whole goat's milk for the first three or four weeks of her life. Cow's milk is not ideal for rearing kids. If she is going to form the nucleus of another herd her new owners will probably have to feed her on milk substitute. It is best that she is gradually weaned on to this by you before she leaves. It will take ten to fourteen days to do it properly.

Always use a good quality orphan lamb substitute and start by mixing up only 1 pt. (575ml) and adding 1 tablespoonful only to her goat milk feed. Store the remainder of the pint in the refrigerator until her next feed. On this occasion add two tablespoonsful of substitute to the goat milk, and so on, gradually changing her food from all goat's milk to the substitute. Try to be patient and not hurry the change-over and it will pay dividends as the kid will be unlikely to scour if this method is followed. When she leaves you, ensure that her new owners have a supply of the same substitute. Calf milk substitute is not good enough. If this is used the kids scour, always seem hungry and tend not to grow so well.

Bottle feeding – note the position of the thumb acting as valve in conjunction with small hole in the teat

The only disadvantage of the orphan lamb substitute is that the manufacturers are catering for sheep and most of the farmers have finished feeding this by the end of June or beginning of July, so a supply can be difficult to locate after that time. Having been warned of this, hold some in stock. Kids will be drinking substitute well beyond the end of June.

All milk should be fed at blood heat. Test by shaking a few drops on to your wrist. At first the quantity the kid takes will fluctuate from 2-10 oz (50-275g) but towards the end of a month she will be happy to consume $1\frac{1}{4}$ pt (725ml) four times each day. Keep to that quantity only. If possible, offer her five feeds each day for the first week but then establish a routine of four milk feeds per day until she is ten or twelve weeks old. By this time she will be eating other things so one milk feed may be dropped. Drop further feeds at approximately monthly intervals until she is having just one each day. Continue giving her one milk feed each day, even if it is not as much as $1\frac{1}{4}$ pt (725ml) until about Christmas time.

If it is easier she can have her milk from a bucket by the time she is three months old, but do not rush into this. When kids drink milk from a bucket they also gulp in air and can get blown. This is distressing to them and the onlooker especially as it could have been avoided. Never let kids suck your finger, for the same reason. Discourage visitors from allowing them to do it too. Feeding the kid more than $1\frac{1}{4}$ pt (725ml) of milk per feed can also make her blown. Encourage the kids to drink warm water from a bucket but take away the bucket afterwards until you are sure she is big enough to come to no harm if she jumps in it when playing.

Bottle-feeding

Kids that are to be bottle-reared should be removed from the dam on the fourth night, after the dam has fed them, and be put in a strong wooden box on a bed of straw or hay. Later on, this box may be turned on its side and used to play on as well as to sleep in. Put the box to the back of their new pen, away from the door and out of the draught. When the kids are too big to sleep completely in the box, they will continue to sleep with their heads in it, because it represents security to them. The dam and the kids will sleep quite happily apart and by the following morning the kids should be hungry. Start by using a small bottle as the kid is unlikely to drink very much and a small bottle is easier to hold in your hand. Use a soft, lamb teat, making sure the hole is big enough. When checking this, make another hole where the teat folds over the rim of the bottle if the teat was bent flat. If you keep your thumb lightly resting on this hole during feeding, it will act as a valve and the teat should not annoy the kid by going flat and, in effect, sealing itself.

When all is prepared, sit the kid quietly on your lap with its head facing your right hand, if you are right-handed. As you would for drenching, put your left arm across the kid's back, cup its chin in your left hand and gently press down its lower jaw with your thumb and insert the teat on top of the tongue. Think how far the dam's teat has been in the kid's mouth and insert the teat of the bottle the same amount. Keep your left hand under the kid's chin at first as you may have to teach the kid to suck the teat by pumping the milk into its mouth with the thumb and index finger of this hand.

If the kid is not very enthusiastic at first about this new method of feeding, put her back in her box and try again in four hours. Some milk is bound to have trickled down and she will come to no harm if she goes without for a few hours. If she appears to have had nothing by the time of her last feed of the day, put a little honey or golden syrup on the teat. She will enjoy licking this and will usually suck from then on.

The introduction of a lamb bar is a good thing when the herd includes six or more kids. Up to twelve kids can be fed on one of these in approximately the time it takes to feed one on the bottle. Equipment for making a multi-feeder of this type should be available locally. Start by drilling up to a dozen holes just below the rim of a $2\frac{1}{4}$-gallon (10l) bucket, and insert the special wide rimmed teats which will not pull through the holes. From the inside of the bucket insert 12 inches (30cm) of plastic petrol pipe-size tubing in each teat. This length should reach from the teat to the bottom of the bucket. With reasonable care this simple piece of equipment will last many years before it needs to be replaced and will save hours of working time. It is easy to sterilize, too.

Male Kids

Also, on the fourth day, you must have decided what is to be done with the unwanted male kids. It is unkind to be sentimental over them. Even if they are pedigree animals, are they really good enough for breeding? Except in very special circumstances, is it wise to keep one for your own breeding purposes? He is possibly already closely related to 50 per cent of the herd so another male will have to be used for them anyway. It is never a good idea to sell or give an unwanted male as a pet to someone else. People forget that unless he has been castrated he will smell during each rut, he is not a lawn-mower, and he will need lots of time and patience spent on him to train him, as would any big dog. The first one, or even two, can be useful if trained to pull a small cart but, being realistic, it is just not economically possible to keep more than that on any holding. If you do not wish to rear for meat, it is far kinder to get the vet to put them down on the fourth day. The dam will not be at all distressed if it is done then, although she will be if it is done earlier.

Unless a male kid has been trained correctly he can be a menace when the rut begins and it is often about this time one starts to hear of tales of neglect. Just prior to the rut is, therefore, the ideal time to get the unwanted males slaughtered for the freezer. Choose an abbatoir that uses the electric stunner first. This has no ill-effect on the animal whatsoever. Particularly if more than one animal is slaughtered at any one time, be wise and joint the carcases yourself. Alternatively, get the slaughterhouse to halve each carcase before you joint it.

If an animal has to be slaughtered, never give him a name, and make up your mind before he is born to send him to the abbatoir. You will not send him there if you do give him a name, but will regret it every time he causes trouble. The most economic time for slaughter is at between fourteen and sixteen weeks of age. After this time the rate of growth decreases in ratio to the intake of food.

Between the tenth and fourteenth day after they have been born, let the kids have access to some earth. They can either eat it from a container placed in their pens, which should be earth from two spits or spade depths down, or they can scrape the loose earth from the roots of grass tussocks. If a container is used, the earth should be renewed each day as the kids will sit in the container after they have eaten their fill. Offer the earth to them until they lose interest, which will probably be in about three weeks' time. The earth from molehills is ideal to use as it is not in a clod and only a healthy mole will throw up a hill. From the earth the kids get a lot of minerals, such as copper, but in the minute quantities they can absorb.

Foraging

By the time the kids are a week old it is noticeable that they are beginning to forage for themselves, although hardly anything is actually eaten. Encourage them to forage by offering bunches of small leaved branches. Ash is a favourite right from the start of their lives. The more roughage they have, the better.

Kids should be introduced to concentrates very gradually. Get a lamb weaner pellet for them first and feed according to the instructions. Follow this with a lamb weaner pencil, which is slightly larger in size, and so on to the same type of food as the milkers get, all the time ensuring that all changes in their diet are made gradually. As a guide, when she is six months old she should be receiving 8 oz (225g) of concentrates per day, split into two feeds, but no more. Keep to this amount until you have to start steaming-up prior to kidding. Roughage and controlled quantities of milk are the important things during the early months.

8

KID MANAGEMENT

Always speak and handle kids quiety and gently. Let them come to you rather than you advancing to them. If they jump up, gently take hold of the front legs, place them on the floor and say 'No'. Only pat or stroke the kids when all four feet are on the ground. They will soon learn what is acceptable behaviour. Teach them their names. Start by saying the name whenever each kid is fed so that the name is associated with something pleasant happening. Eventually you will be able to call each goat individually from the herd.

Castration

Non-breeding males that are to be kept through the months of the rut should be castrated. There are two methods commonly practised – ringing and cutting. A kit for ringing is available on the market but it is advisable to watch this being done by someone else before attempting to do it yourself. The second day of life is the recommended time to do it, before the scrotum is very big. Do be absolutely sure that both testicles are down in the bag before attempting to place the ring in position. Paralysis of one or both back legs can result if the instructions are not followed carefully. The kids always appear to suffer considerable discomfort for at least an hour with this method because the supply of blood has been cut off from the scrotum.

Both the ring and the shrivelled scrotum will drop off after about seven weeks.

Because of the number and variety of animals vets have to deal with, they are very quick and efficient at castrating the kids by cutting, and if the kid is taken to the surgery the cost is comparatively low. Even in high summer there is very little risk of infection although occasionally there might be a slight haemorrhage in one groin. It is only likely to occur once or twice in the whole of a goatkeeping career, but check that everything is satisfactory if he is not chasing around with the others. If a kid does have a haemorrhage, separate him from the others so that he does not get trodden on. If he is left quietly on his own, and the feeding routine is still maintained, the swelling will disappear within twenty-four hours. No bathing or medication at all is required, just peace and quiet.

Disbudding

There are several methods of doing this but the most common one is to use a hot iron. A male kid that is to be kept for breeding should have his horn buds removed when he is four days old. It is then only a minor operation but after the fourth day the horns seem to grow so fast it rapidly becomes a major one. Try to get the disbudding done before the kid is finally taken away from his dam and then he can have a comforting suck as soon as he gets back from the vet. They both enjoy this and are satisfied that all is well with their world.

Most vets use a local anaesthetic, giving the kids four injections, one in front of each ear and one behind. Kids have four nerves in this area, unlike calves who have two. A few vets, and some very experienced goatkeepers, who will sometimes do disbudding for fellow club members, give a general anaesthetic. This does allow more time but not all goats can cope with a general anaesthetic and some cannot then be revived. This is rare nowadays but one should be prepared for it to happen.

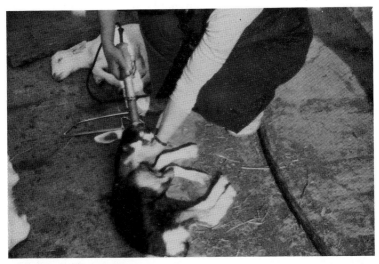

Disbudding with a hot iron

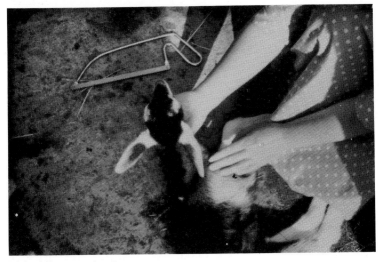

Disbudding completed

Another rare happening which can be quite frightening at the time if you are not prepared is a kid that goes stiff when it has its injections. The disbudding can still take place but on returning home put the kid in a separate pen, well padded with bales of hay or straw, and try to ensure that someome stays with it until it is firmly on its feet. A sensible child can stay with it for a couple of hours; it does not have to be an adult.

The horn buds of female kids grow more slowly and provided they are no bigger than your little finger nail they may be left until the kid is ten days old.

It is advisable to let the vet know beforehand that you are bringing a kid for disbudding or, in these days of partnerships, you may arrive at the surgery only to be told that all the disbudding irons are with the other vets, on the farms. A wasted journey is both time-consuming and uneconomical.

As soon as a kid is born it is possible to tell whether it will have horns. Dampen the top of the kids head and, if the fur goes into a whirl, horn buds are due to appear within the next day or two. Otherwise the kid will look as if it has a fringe.

An alternative to the disbudding iron is caustic pencil but in inexperienced hands the horn continues to grow and, in most cases, this is then malformed. Use caution with this method.

Some Words of Warning and Reassurance

The Warning. Kids are very active youngsters so give them something on which to play. An old tree trunk is a wonderful thing for this if it is available. As an alternative, four or five bales of straw make a good jump-off for a game. Put their strong wooden box in the yard too, from time to time, to add variety. Because kids can, and do, jump so high make sure there is never a loop of string anywhere near their habitat. Only use knots, not bows, to tie up branches and so on. More legs have been broken and kids hung because of loops thoughtlessly left around than for any other reason.

The Reassurance. When a kid is tiny it can give a beginner, and some more experienced goatkeepers too, a fright when it sleeps. The breathing is very shallow then. Couple this with the fact that many kids will sleep in a corner, several feet away from the dam, and one immediately begins to suspect it is dead. Be reassured – this is most unlikely to be the case. If it was going to die you would have noticed something wrong long before this.

Registration and Transfer of Ownership

Any female kid which has had a pedigree sire may be registered after it is one month old. Male kids may only be registered if both parents are pedigree animals. All kids that are to be registered from one litter must be registered at the same time. Registrations may be made through the Secretary of your local affiliated club, who will countersign the form, or direct with the British Goat Society if you are a member. Registration forms are obtainable from club secretaries and many Society members free of charge.

Transfers

Whenever a kid, or adult, is to be bought the transfer of ownership should be noted on the registration card. Forms to enable one to do this are obtainable from club secretaries and the British Goat Society.

9

BREEDING PROBLEMS, ILLNESS AND OTHER WORRIES

False Pregnancy

The reason for a false pregnancy, or Cloudburst as it is more often called, is still not fully understood but it is believed to be psychological. The goat looks very rotund and appears to be in kid but you know that she has not been to the male. Her udder does not change in size, as it would if she was in kid, as the unknown kidding date approaches. Eventually she does 'kid' but, of course, there is no kid, only pints and pints of what looks like water appears, hence the name Cloudburst. When this happens just give her a dry bed, carry on milking as usual, watch her extra carefully for a few days and put her to the male as soon as conveniently possible. She is unlikely to come into season again until the following autumn.

Abortion

This is rare but when it occurs six weeks after conception it is probably caused by a vitamin A deficiency. If it happens nearer kidding time, it could have been caused by a swinging gate banging into the goat, or one of the other goats butting her. Alternatively, she could have repeatedly had to pass through a doorway that was too narrow. In all cases, check that the afterbirth has come away and review your system of management.

Where young stock has died as a result of coccidiosis, a vitamin A deficiency could occur in the adults.

Reabsorption

Here, the pregnancy appears to be normal but the goat's udder never floods and she never kids. A few days after the anticipated kidding date you will no doubt check, and re-check, the kidding date or ask the vet to inspect her. You will be told there is a kid there but still she produces nothing. Gradually she will get less and less rotund and perhaps you will see a small, mouse-size, object, or a slight red/brown discharge. This could happen as much as three months after the true kidding date. The cause is a vitamin E deficiency. Give her a teaspoonful of wheatgerm in each feed of concentrates for a few weeks prior to getting her into kid early in the next rut.

In fact, any goat that has had difficulty in holding to service one season could be fed a small quantity of wheatgerm for a week or two prior to her next mating. It certainly will have no harmful effects on her if she does not need it and could make the difference between one or several trips to the male.

Hermaphrodites

Hermaphroditism is not common these days but it is closely linked to the gene factor in naturally hornless goats.

Hermaphrodite kids should be put down at four days or reared for meat only. They are no good whatsoever for breeding. In most cases, as soon as a kid is born you will notice a small blob of skin at the point of the vulva, like a pea, and this could get progressively larger as the kid grows until it is about the size of a grape on a stalk.

To avoid hermaphroditism always ensure that either the dam or the sire of a future kid have themselves been born with horns. So long as one of the pair have had horns the kid should be normal in this respect.

Maiden Milkers

A few goats start to develop an udder during their second summer. These are known as maiden milkers and will produce milk before they are mated. During the first few days a maiden is milked she will only produce a thin, pale brown liquid but by the end of the first week it will begin to look and taste like milk. Some maiden milkers will gradually increase their yield up to 2 or 3 pints (1 or 1.5 l) per day if they are milked at regular intervals and fed accordingly, but if they are discouraged and only milked when they are uncomfortable they will probably milk better after they are mated. It depends what you want from your goat as to which method you use.

If an udder develops in a kid, advice should be sought from the vet.

The Medicine Chest
A biscuit tin is large enough to contain the average beginner goatkeeper's medicines. Such things as cotton wool, strips of old, clean sheet or bandages of varying widths, scissors, a dose of worm tablets, some liquid paraffin or gripe water and antiseptic udder cream should be kept ready for an emergency. Try to make a back-to-front coat from a hessian sack. Just put two holes in it for her front legs to go though and add four or six tapes to tie it in position along her back. This can either be used as a lightweight rug to keep her clean for a short period on a special occasion or, with her front legs in the holes, it will stop her nibbling at any bandage she may have across her chest. All tapes should be tied with knots, and pins or bows kept well away from all the goats.

Keep the drenching jug, or small teapot in the first aid kit, as well as a 4-foot (1.3m) piece of washing line rope, in case you have a difficult kidding.

As you become more experienced at coping with illness and accident, a syringe, the white spirit to clean it, terramyacin, and so on will be added but keep things simple at the beginning. The normal temperature of a goat is 102.5° to 103°F (39°-39.5°C) and the thermometer is inserted into the rectum.

Fractures

These can be dealt with quite successfully by your vet but where a young animal is involved the plaster should be left on for no more than three or four weeks. Young animals grow so fast that plaster casts left on beyond this time are very difficult to remove. Should the cast need to be in position longer, get the original removed and have it replaced with another.

Shock can have a worse effect on the goat than the fracture, so see that she has access to plenty of fibrous roughage. She will quickly learn to cope with the weight of the plaster on her leg.

It is possible to mend a broken leg

Spring Fever

In some years winter is so bad that the goats can go out very little and this leads to what is called Spring Fever. This is not a fever; it is scurvy. The goat's fur feels greasy and is full of loose flakes of skin. Goats will often pull out tufts of their own fur, drawing blood in so doing, but the condition will clear once they can get some fresh greenstuff into them again. Try to avoid it happening by feeding bunches of hand-picked fresh roughage. It is also a help if at the first signs of the condition quarter of a 50mg vitamin C tablet is crushed and sprinkled on the concentrate ration. Gradually increase this quantity to 25mg or 50mg per day, depending on the size of the goat.

Mastitis

Dirt is the cause of this. Even tiny scratches on the udder should have an antiseptic cream applied to them. A cream containing lanolin is helpful as it is absorbed quickly. Where mastitis is present lumps will be felt in the udder and the milk from the affected quarter will look like string. Internal bruising caused by a bang or a fall, or rough handling during milking will in time lead to mastitis. A wet or damp bed will certainly be a factor in encouraging the infection to spread.

With careful management the goats will get very little wrong with them beyond the occasional cut or scratch, but sharp eyes, alert ears and gentle hands will anticipate and so avoid most troubles. Barbed wire and old sheets of corrugated iron used to fill gaps in hedges should be removed from all holdings where there are goats as soon as possible and be replaced with something more suitable. This is not always easy but it is worth the effort.

During any illness the attendant's attitude is crucial to the well-being of the patient. However despondent you feel you must always speak cheerfully to the goat or the despondency will be communicated. Sometimes it is necessary to put her in another pen but, unless there is a special reason for not doing so, put her where she can see and hear the others. This will help to hold her attention and keep her alert. Keep the routine as near the normal one as possible so that she thinks it is normal. Goats that are ill are at their lowest ebb at about 2 a.m. It will therefore be an advantage to try and give her a visit about that time to check that she is warm enough. Encourage her to eat or drink a little to help to keep her warm.

Spring fever – Scurvy

Routine Medication

Worming. Unless any indication is given that a particular goat should have a further dose, worming only takes place in the spring (before the goats start to go out regularly to graze, and avoiding the weeks prior to kidding and immediately afterwards), and in the autumn, prior to the rut.

Where goats are on free-range it discourages the reinfestation of worms if their paddock can be divided into four smaller ones.

The life cycle of the common stomach worm is three weeks. If the goats are rotated around the smaller paddocks over a four-week period, spending the first week in paddock 1, the second week in paddock 2, and so on by the time they return to paddock 1 this will have had a full month free from the goats and be clean again.

Fluke

Fluke is prevalent in certain wet areas, and as the snail whose life is an intrinsic part of the fluke cycle can travel up to 2 miles, any wet area in the locality is suspect. In live animals fluke is very difficult to define as its presence is insidious. The animal just does not do well.

If fluke is known to be in an area it is a wise precaution to drench for this on a routine basis. Do it in the spring and autumn, as for worms. Some medications will deal with both worms and fluke at the same time.

10

SHOWING

Whilst showing goats is not everyone's ideal afternoon's entertainment, it must be remembered that shows are every goatkeeper's shop window, and they are often many people's first and only contact with goats. Much can be learned from talking with others with a similar interest. Why not volunteer to help on a publicity stand as an introduction to showing?

Even quite small clubs hold a members show once each year and these are great fun, with classes for the children as well as the adults and their goats. In fact, when it comes to the child handler classes, a lot of the children put the adults to shame. Many of the big agricultural society shows try to find time to include a child handler class when the advertised classes are over. Club shows give one an opportunity to start learning the business of showing. The vast majority of goatkeepers are only too pleased to help beginners. Even if a goatkeeper has only entered one or two shows previously he should have learned something to pass on to others.

Goats should be well groomed and have had their feet trimmed a couple of days before a show. No grooming equipment should be used in the ring and certainly the judge should not be asked to 'Hold on a minute' whilst feet have their last trim. You should aim to be able to take your goats from their home pen, giving them a quick brush to remove any odd bits of hay, and straight into the show ring. They should not need washing or to have talc and chalk rubbed into their white coats, although some goatkeepers do like to trim their beards beforehand. Horns do not disqualify a goat from being entered in a show.

Child handlers at a club members' show

Club show secretaries often say they will accept entries on the day of the show, on the field, but all B.G.S. recognized shows and agricultural society shows have a closing date for entries to which one must adhere. About three months before a show date is a good time to contact the secretary and ask that your name and address be added to the list of people to whom schedules will be sent. Subsequently, the schedule will be sent to you automatically.

Milking Trials

Prior to a show with milking trials, where you will have to stay overnight, see that all equipment is marked with your name or some distinctive sign to make it easier for the show secretary to identify anything you may leave behind. Try to take a white overall with you, not forgetting a clothes peg or safety pin to attach the goat's number cards to a top pocket. The goats like a change in their diet and find number cards very palatable! Judges, understandably, do not always appreciate this habit. Bedding, hay and branches, as well as concentrates, must be taken to milking trial shows. Only a comparatively few branches or hay are needed for club events as the goats are frequently left in their travelling vehicle, which should be parked in the shade, or put on a lead and walked around the field with their owners when they are not in the ring.

Arrive at a show in good time, check which pens you have been allocated and get the goats into them. Give them a feed of hay and a drink of warm water. Some show secretaries arrange for hot water to be available in the marquee. If a wash-boiler is used for this, remember to put in some more cold water to replace the hot you have just removed. When the goats are settled – and some kid owners find it an advantage to take a hurdle or small meshed net to place on top of their pen to stop them jumping out – your own needs may be attended to. Concentrates should not be fed immediately after a journey. Leave this for at least an hour after arrival.

On the first evening, all the milkers will be stripped out under the judges' or stewards' eyes at a specific time, probably 6 p.m., regardless of the time of their previous milking. The next two milkings, on the following day, also take place before the judge and stewards, and these are the ones that will be weighed and tested for butterfats so be prepared to stay on the showground until after the second milking, however long your journey home may be.

Procedure in the Ring

Whether it is a club event or something bigger, get your goat ready to enter the ring with its collar and lead on as soon as the steward tells you. Walk your goat quietly and quickly into the ring, stand her well, and concentrate on the judge and your goat only whilst in the ring. At the end of each class, if time permits, the judge will tell you why you have been placed in your particular position. Learn from this. If you are given a rosette and small prize 'Hurrah', but if not have another go next year. Goatlings are very much like teenagers and go through gangling and awkward phases, and even if she was not placed first, second or third you will have got some experience of a show routine and atmosphere. Make use of this on another occasion. Next time you will be in the position to help another less experienced than yourself.

Finally, be prepared to take your goat into the grand parade.

USEFUL ADDRESSES

Clubs and Societies

Anglo Nubian Breed Society, Hon. Sec.: Miss N. Wing, Hall Farm Cottage, Hooton Lane, Ravenfield, Rotherham S65 4NH.

Avon County Goat Club, Hon. Sec.: Miss J.G. Macleod, Chestnut Tree Cottage, Doynton, Bristol BS15 5TA.

Bedfordshire & Hertfordshire Goat Society, Hon. Sec.: A.F. Munnings, 276 Mancroft Rd., Ally Green, Luton, Beds. LU1 4RD.

Berkshire & Oxon Goatkeepers Federation, Hon. Sec.: C.W. Hawley, Watersmeet, Sheffield Bottom, Theale, Berks.

Bingley & District Goat Club, Hon. Sec.: Mrs D. Wright, Bay of Biscay Farm, Haworth Road, Allerton, Bradford, Yorkshire.

Buckinghamshire Goat Society, Hon. Sec.: Mrs Nicholson, North Park Lodge, Chequers Estate, Butlers Cross, Aylesbury, Bucks.

Cambridgeshire & District Goat Society, Hon. Sec.: Mrs C. Brook, Keeble Cottage, 101 Longstanton Rd., Oakington, Cambs.

Cheshire Dairy Goat Society, Hon. Sec.: Miss J. Vile, Ryton House Farm, Wrenbury Heath Road, Sound, Nantwich, Cheshire.

Cheshire Goatkeepers' Association, Hon. Sec.: Mrs B. Pierce, 142 Park Rd., Great Sankey, Warrington, Cheshire.

Cleveland Dairy Goat Society, Hon. Sec.: Mrs L. Weedy, Oaklands Farm, High Leven, Yarm, Cleveland.

Colchester & Sudbury District Goat Club, Hon. Sec.: Miss P.V. Minter, Avon, Park Lane, Langham, Colchester, Essex CO4 5NJ.

Cornwall Dairy Goatkeepers' Association, Hon. Sec.: Mrs E. Morris, Marroy, Toldish, Indian Queens, St Columb., Cornwall TR9 6HL.

Derbyshire Goat Club, Hon Sec.: Mrs W. King, 24 Bostocks Lane, Risley, Derbyshire.

Devon Goat Society, Hon. Sec.: Mrs M. Smith, 27 Stenhill, Uffculme, Cullompton, Devon.

Dorset Goat Club, Hon Sec.: Mrs M. Bentley, 9 Bladen Valley, Briantspuddle, Dorchester, Dorset DT2 7HP.

Durham Dairy Goat Society, Hon. Secs.: Mr and Mrs M. Harbour, The Bungalow, Golden Corner, Byers Green, Spennymore, Co. Durham DL16 7NW.

English Golden Guernsey Goat Club, Hon. Sec.: Esme Brown, East Johnstone, Bish Mill, South Molton, Devon.

Fenland Goatkeepers' Society, Hon. Sec.: Mrs Y. Eastgate, Craven Cottage, Croft Road, Upwell, Wisbech, Cambs.

Frome & District Goatkeepers' Club, Hon. Sec.: Mrs P. Hogg, Silver Street Farm, Brokerswood. Westbury, Wiltshire.

Glamorgan Goat Club, Hon. Sec.: J.E. Evans, Tredegain Farm, Pencoedcae Road, Beddau, Pontypridd, Mid. Glam.

Gloucestershire Goat Society, Hon. Sec.: Mrs A.B. Benn, Mill House, Framilode, Saul, Gloucestershire.

Grampian Goat Club, Hon. Sec.: Mrs C. Kennedy, Jimpies, Culsalmond, By Insch, Aberdeenshire, Scotland.

Hampshire Goat Club, Hon. Sec.: Mrs B.M. Bull, Beacon Cottage, Beacon Bottom, Swanick, Southampton.

Helmsley & Kirkbymoorside Goat Club, Hon. Sec.: Mrs B. Hughes, Howkeld Cottages, Kirkbymoorside, York YO6 6HB.

Herefordshire Goat Club, Hon. Sec.: Mrs J.E.M. Winter, The Old Farm House, Leinthall Starkes, Ludlow, Salop.

Highland Goat Club, Hon. Sec.: Mrs E. Bishop, 44 Bason Road, Inverness, Scotland.

Hull & East Riding Goat Society, Hon. Sec.: Mrs R.M. Mylme, Kirkdale House, Huggate, York Y04 2YF.

Ipswich & District Goat Club, Hon. Sec.: Mrs C. Gosling, Rosanda, Knodishall, Saxmundham, Suffolk IP17 1UA.

Irish Dairy Goat Society, Hon. Sec.: Mrs A. Bridle-Parker, Kracken House, Ferefad, Longford, Eire.

Irish Goat Club, Hon. Sec.: Miss A. Kidman, Ballymacad Stud, Oldcastle, Co. Meath, Eire.

Kent Goat Club, Hon. Sec.: Mrs S. Whatman, Thornton, Mayfields Lane, Wadhurst, Sussex.

Lancashire Dairy Goat Society, Hon. Sec.: Mrs J. Marriott, Mount Pleasant Farm, Haslingden Old Road, Oswaldtwistle, Lancs.

Lincolnshire Goat Society, Hon. Sec.: Mrs V. Wilmot, Happylands Farm, Owersby Moor, Market Rasen, Lincs. LN8 3YN.

Mid Essex Goat Club, Hon. Secs.: Mrs & Miss Trigg, The Anchorage, Nathans Lane, Writtle, Nr. Chelmsford, Essex.

Mid Wales & Border Goat Society, Hon. Sec.: Mrs B.E. Evans, Waen Uchaf, Bwlch-y-cibau, Llanfullin, Powys.

Norfolk & Suffolk Goat Society, Hon. Sec.: Mrs F.R. Buck, Pollard Tree Farm House, Wortham Ling, Diss, Norfolk.

Northants & District Goat Society, Hon. Sec.: D.E. Brace, Victoria House, Hilton Rd., Fenstanton, Huntingdon PE18 9LJ.

Northern England Goat Club., Hon. Sec.: V.J. Riley, Sunnymede, Port Carlisle, Carlisle, Cumbria CA5 5BU.

Northern Ireland Goat Club, Hon. Sec.: Miss Barbara Gibbins, Finlarig, Burren, Ballynahinch, Co. Down, N. Ireland.

North Leicestershire Goat Club, Hon. Sec.: Mrs W.V. Turner, 71 High St., Chellaston, Derbyshire DE7 1TB.

North Staffordshire Goat Club, Hon. Sec.: M.L. Downing, East View Cottage, Grindon, Nr. Leek, Staffs.

North Wales Goat Society, Hon. Sec.: M.J. Colborn, Maes Hyfryd, Station Road, Llanrug, Caernarfon, Gwynedd, Wales.

Northumbrian Dairy Goat Society, Secretary: Mrs Jill Harrison, Morrelhirst, Netherwitton, Morpeth, Northumberland NE61 4PT.

Norwich & District Goat Club, Hon. Sec.: Mrs M. McDonald, Ivy Cottage, The Hill, Antingham, North Walsham, Norfolk NR28 0NH.

Nottinghamshire Goat Club, Hon. Sec.: Mr C.B. Marrison, 8 Sheppards Row, Queen Street, Southwell, Notts.

Orkney West Mainland Goat Society, Hon. Sec.: Mrs Anderson, Kelda, Twatt, Orkney KW17 2JD.

Pennine Dairy Goat Club, Hon. Sec.: Mrs J. Mynard, 9 Mill Hill, Brearley, Luddendfoot, Halifax, West Yorkshire.

Pontefract & District Goat Club, Hon. Sec.: Mrs B.A. Caswell, Moorgate House, Hungerhill, Boisterstone, Sheffield S30 5ZF.

Rydale & District Goat Society, Hon. Sec.: Mrs M. Braithwaite, Lunds Garth Farm, Middleton, Pickering, Yorks.

Scottish Goatkeepers' Federation, Hon. Sec.: Mrs E. Mitchell, Conachra Farm, Gartocharn, Dunbartonshire.

Shropshire Goatkeepers' Society, Hon. Sec.: Mrs E. Burt, White Cottage, Golden Grove, Llysty, Bishops Castle, Salop.

Somerset Dairy Goat Club, Hon. Sec.: Mrs B. Franche, Orchard House, Hurst, Martock, Somerset TA12 6JT.

Southern Scotland Goat Society, Hon. Sec.: Miss K. Wakefield-Richmond, St Michaels, Dryfesdale, Lockerbie, Dumfriesshire.

South Leicestershire Goat Society, Hon. Sec.: Mrs C.J. Brown, Cherry Tree House, Dunton Basset, Lutterworth, Leics.

South West Wales Goat Club, Hon. Sec.: I. Grant, Green Bush, Clynderwen, Dyfed.

Surrey Goat Club, Hon. Sec.: Mrs H. Hanbury, Foxbury, The Ridge, Woldingham, Surrey.

Sussex County Goat Club, Hon. Sec.: Mrs Paris, Willow Cottage, East Chiltington, Nr Lewes, Sussex.

Swaffham & District Goat Club, Hon. Sec.: A.E. Hall, Hollow Dene, Little Ryburgh, Fakenham, Norfolk NR21 0LS.

Warwickshire Goat Society, Hon. Sec.: Mrs R. Ragg, Foxcote Farm House, Shipston-on-Stour, Warwickshire CV36 4JG.

Waveney Valley Goat Club, Hon. Sec.: Mrs P. Carter, Blue Tile Farm, Brampton, Beccles, Suffolk.

Welsh & Marches Goat Society, Hon. Sec.: Mrs L. Partridge, Rosedale Stud, Gelli Farm, Henllys, Cwmbran, Gwent.

Wessex Goat Club, Hon. Sec.: Mrs J. Ballam, Eastwood, Danes Hill, Dalwood, Nr Axminster, Devon EX13 7HH.

West Midlands Goat Society, Hon. Sec.: Mrs J.F. Lasance, Invicta, Bishopswood, Brewood, Staffs. ST19 9AD.

Wiltshire Goat Society, Hon. Sec.: Mrs E.M. Anderson, Forli Acres, Minety, Malmesbury, Wilts. SN16 9RQ.

Worcestershire Goat Society, Hon. Sec. Mrs C.M. Wickett, The Orchards, Harvington, Nr Evesham, Worcs.

Wye Valley Goat Club, Hon. Sec.: Mrs L. Gwyn, Marlborough House, Newton St Margarets, Vowchurch, Herefordshire.

Yorkshire Goat Society, Hon. Sec.: Mrs Harben Williams, Field House, Gomersal, Cleckheaton, Yorks.

Overseas Societies

Tobago Dairy Goat Society, Scarborough, Tobago, West Indies.

New Zealand Goat Breeders' Association, Hon. Sec.: B. McKerras, 87 School Road, Mai Etal Beach, Auckland, New Zealand.

Croydon & District Goatkeepers Club, Dalkeith, Main Hurstbridge Road, Wattle Glen, Victoria 3096, Australia.

Mevr. v. Kranenburg, Klutenpad 17, Creil (N.o.P.), Holland.

American Goat Society, Inc., Mr J. Willett Taylor, 1606 Colorado Street, Manhattan, Kansas 66502.

Parent Associations

The British Goat Society, Rougham, Bury St Edmunds, Suffolk
   IP30 9LJ.

Scottish Goatkeepers' Federation, Hon. Sec.: Mrs E. Mitchell,
   Conachra Farm, Gartocharn, Dunbartonshire.

Mevr. v. Kranenburg, Klutenpad 17, Creil (N.o.P.), Holland.

Just Plain Useful

Henry Doubleday Research Association, Convent Lane,
   Bocking, Braintree, Essex.

W.W.O.O.F. (Working Weekends on Organic Farms), 56, High
   Street, Lewes, Sussex BN7 1XE.

National Centre for Alternative Technology, Machynlleth,
   Powys, Wales.

British Denkavit Ltd, P.O. Box 6, Agrarian House, Castle
   Street, Poole, Dorset BH15 1HL. (Goat and Kid Food)

Sea Products Research Ltd, Pound Batch Farm, Abbots Leigh,
   Bristol 8. (Seaweed Meal)

O.F.I.C. (G.B.) Ltd, William Street, Northam, Southampton
   SO1 1QH. (Roofing Sheets)

Rossendale Electronic Fencers, Lumb-in-Rossendale,
   Lancs. BB4 9NJ

Brewers' Grains Marketing, Wetmore Rd., Burton-on-Trent,
   Staffs. DE14 1TF

Lakeland Plastics, Alexandra Buildings, Station Precinct,
   Windermere, Cumbria LA23 1BQ.

Nutrichip (Goat Mineral Chips), Corsham, Wiltshire.

INDEX